All About Apples

Adam and Eve (The Granger Collection, New York)

ALICE A. MARTIN

All
About
Apples

with illustrations

HOUGHTON MIFFLIN COMPANY BOSTON
1976

Library of Congress Cataloging in Publication Data

Martin, Alice A
  All about apples.

  Includes index.
  1. Apple. I. Title.
  SB363.M28   634'.11   75-32404
  ISBN 0-395-20724-X

Printed in the United States of America

C 10 9 8 7 6 5 4 3 2 1

My heart is like an apple tree
Whose boughs are bent with thick-set fruit.

D. G. Rossetti, "A Birthday"

Acknowledgments

MY THANKS to "Farmer Carr" and Mrs. Norris Johnson for their personal accounts of apple growing and to Russel Allen of Bellows Falls, Vermont, for current material he kindly lent me. Mr. and Mrs. Ken Perry of Williamsville, Vermont, were zealous in their search for nineteenth-century agricultural books and periodicals to add to my collection.

To Frances Tenenbaum of Houghton Mifflin my thanks and affection, which feelings antedate this book by many years. And to my family, but especially J.M. and L.M., my thanks for their help and forbearance.

Contents

Illustrations

following page 80

In a cider-mill.

The old "horse motive power" cider press, 1905.

Apple trees in winter.

Spraying an apple orchard in Douglas, Michigan, with
a power sprayer on a horse-drawn rig, 1915

Grading and barreling apples, 1915

A roadside farm market, 1920

Paring apples for a pie.

An early twentieth-century apple-bobbing party.

In the early 1930s the apple became a symbol of the
Depression.

(*All illustrations except "In a cider-mill" and "Apple
Trees in Winter" are by Brown Brothers.*)

Here's to thee, old apple tree,
Whence thou mayst bud, and whence thou mayst blow;
And whence thou mayst bear apples enow.

Old English farmer's wassail
at harvest-time

Introduction: Before the Apple, the Tree

THE APPLE TREE: neither the magnificence of the oak, nor the grandeur of the beech or maple. The singularity of *Pyrus malus* among the great trees of the earth lies in its relation to man. The apple is the most domestic of trees and more than any other symbolizes the real and wished-for virtues of Home. It is not a tree of the forest, although it can survive there; its natural habitat is by the side of man. Its fruit has been his food for millennia.

The tree is sturdy, yet hardly forbidding. The trunk is short with branching limbs, and its overall height is compatible with man's dwelling. It brings forth fruit abundantly; it is rugged and able to survive the vagaries of climate and soil, sometimes for many years past man's own span. For all these reasons the fruit and the tree have been cherished wherever in the temperate zone man has settled.

At any season of the year it is good to look at. The winter tree is bold and strong. The denuded branches, even when thriftily shaped to the orchardist's needs, grow all akimbo, in dramatic contrast to the straight-growing maples and ash. On a hillside its silhouette can be easily made out, its limbs at all angles against the sky. After a stormy day the snow layers its blackish bark, outlining its angularity. Here and there the snow or wintry winds weaken a vulnerable limb, and it falls to the ground.

Spring brings the branches tips of green, the first somewhat brownish, and before long new shoots, bearing leaf and blossom-buds, begin to emerge. Then the vision of the flowering tree — white, pink, and green, the lovely apple green — comes into view. With the opening of the blossoms in the warming sun comes the apple scent, faint if from one tree, intoxicating if from a whole orchard. Now the bees are heard, their hum the companion of blossom time.

By midsummer the bees have turned their attention elsewhere, the deep green canopy of leaves almost covers the boughs, and fruit begins to appear. At first hard and green, barely covering the seeds it encases, it swells out in shape, color, and size according to its kind. At the end of the summer the healthy tree, in a good year, seems swollen with fruit, bringing to mind Biblical images of fecundity. The air is still, and only an occasional weakened and imperfect apple falls to the ground. The tree holds the good apples firmly till the maturing fruit is ready-ripe.

In autumn, if left to its own devices, the tree would give up its apples by the process of cutting off moisture to the fruit, weakening the connection with the stem, and "letting-go" of the apple. Man intervenes before this happens and picks the fruit unbruised. So the tree is harvested, the leaves begin to turn yellow and sere, and the tree's life system, no longer concerned with reproduction for that year, begins to go into dormancy. Soon with the wind's help, the tree is winter-bare.

The apple tree has survived another year, a few branches gone, a bit more gnarled, but roots sound, trunk intact, and ready for many more years of life. And with a minimum of care from its "owner" it can go on yielding apples, and serving as part of the miraculous chain of life.

All About Apples

And pluck till time and times are done
The silver apples of the moon,
The golden apples of the sun

W. B. Yeats, "The Song of
Wandering Aengus"

1. The Apple as Symbol: Legends and Lore

LIFE ETERNAL, sexual vigor, fertility, all these and more are powers ascribed to the homely apple at some time or place. From the Bronze Age until the present, one is struck by the variety and pervasiveness of the apple metaphor, whether it represents the female breast or the globe we inhabit. And quite understandably. As the apple slowly grows from nascent bud to bursting plumpness it brings to mind, somehow, a universal burgeoning; in maturity the satisfying roundness of the fruit, so well suited to the hand, seems always to have been a favorite shape, ready-made for so many analogies.

Man has always gone beyond the limiting world of the real to find added levels of meaning. The further back in time, or the simpler the life, the more deep-rooted was the magical power belonging to apples. In remote Kara-Kirghiz, until quite recently, people believed that barren women would be made fertile by rolling on the ground under an apple tree. Similarly, in Oriental Jewish lore, women who wished to conceive were advised to wash their hands with water to which apple juice was added. Such customs have existed wherever apples grow.

The Book of Genesis does not speak of the apple, but rather "of the tree which is forbidden," and some scholars have made a case for the apricot as "the" fruit in the Garden of Eden. The implicit assumption, however, has always been that it was indeed an apple with which the Serpent tempted Eve, and literature, sculpture, and painting abundantly illustrate the belief.

The "knowledge" inherent in the tree and its fruit is often taken to be carnal knowledge, or, at its broadest, the distinction between good and evil. Far more is implied. As it does in other religions, the tree in the Bible represents the knowledge of infallibility and eternal life, possessed by gods but denied to man. Thus, in Genesis: "For God doth know that in the day ye eat thereof, your eyes shall be opened and ye shall be as gods, knowing good and evil." But the delights of the apple, symbolic or otherwise, were too compelling, "And when the woman saw that the tree was good for food, and that it was pleasant to the eyes, and a tree to be desired to make one wise; she took of the fruit thereof, and did eat; and gave also unto her husband with her, and he did eat."

Milton, in *Areopagitica*, gives a somber poetic rendering of this:

> It was from out the rind of one apple tasted
> That the knowledge of good and evil
> As two twins cleaving together
> Leaped forth into the world.

Yet this same story also laid the groundwork for the role of woman as tool or temptress of evil. Thomas Hood, the eighteenth-century poet, writes a beguiling version of this stereotype:

> When Eve upon the first of Men
> The apple pressed with specious cant
> Oh! what a thousand pities then
> That Adam was not Adamant!

In the Song of Songs the apple loses its portentous quality and becomes a frankly erotic symbol. "As the apple tree among the trees of the wood, so is my beloved among the sons . . . Stay me with flagons, comfort me with apples: for I am sick of love . . . Now also thy breasts shall be as clusters of the vine, and the smell of thy nose like apples."

The biblical phrase "the apple of his eye" is found throughout the Old Testament, where it has only one meaning: the regard of the Lord Jehovah for the Israelites. "For he that toucheth you toucheth the apple of his eye," warns Moses in Deuteronomy. In Lamentations Jeremiah gives us a hint of how the phrase, "O wall of the daughter of Zion, let tears run down like a river day and night; give thyself no rest; let not the apple of thine eye cease," evolved. Clearly, what is meant here is the *pupil*, which was identified with apple, a solid sphere. What could be more precious than the apple of one's eye?

This specific use of the apple shape is something of an exception in apple symbolism. Generally, in antiquity, the apple is seen either as an expression of eternal life, or eroticism and fertility. In the tangled stories of classical mythology the apple appears often made of gold, clear measure of the esteem in which it was held. In Greek mythology Mother Earth, Rhea, gave the golden apple tree of eternal life to Hera on her marriage to Zeus. Hera placed the tree on an island at the end of the world, in the safekeeping of Atlas' three daughters, the Hesperides. With the help of the never-sleeping dragon, Ladon, they watched over the symbols of immortality to which only the Gods could have access.

The apples of the Hesperides figured in the adventures of Hercules and Perseus, both by-blows of Zeus, and thus both objects of Hera's jealousy. In other stories, though, the apple is a symbol of Aphrodite, the Goddess of love. When the beautiful — and fleet-footed — maiden Atalanta was told by the oracle that she would meet her death through marriage, she retreated to the woods and announced that any man who wished to marry her would have to beat her in a footrace but those whom *she* defeated would die. One suitor, Hippomenes, knowing he could not win by skill alone, appealed to Aphrodite to favor his cause. Quickly the Goddess gathered three golden apples from her garden and gave them to the lovesick Hippomenes. As the race began and Atalanta gained on her suitor, he threw the first golden apple in her path. Entranced, she stopped to pick it up. When she again nearly overtook Hippomenes, he threw the second one. Again she stopped, and he again drew near. When he threw the third, she was unable or unwilling to recover her lead, and Hippomenes claimed her as his prize. Alas, the fates are not to be gainsaid. Atalanta and Hippomenes were so engrossed in each other they failed to thank the Goddess properly and, passing a temple, entered and made love, thus defiling a holy place. For their outrageous lapse, the two were turned into a lion and lioness, but in honor of their gallantry they live on as stars in the firmament.

The most celebrated apple of myth was that which led inexorably to the long war between the Greeks and the men of Troy. The legend began when Hecuba, Queen of Troy and wife of King Priam, gave birth to a son named Paris, whom the soothsayers declared would be the destruction of his family and city. King Priam and Hecuba gave the child to a shepherd, with instructions to abandon him on Mount Ida, where he would die. The shepherd did not obey his instructions but raised Paris in the simple life of the herdsmen.

Many years later, there was a wedding of great importance. Everyone of note was invited. But even as on mortal guest lists, there was an oversight. Eris, Goddess of Discord, had not been included among the guests. In fury, she threw a golden apple — the Apple of Discord — diabolically inscribed "to the fairest," into the assemblage at the wedding party. A formidable trio claimed the apple and the title of fairest: Hera, Aphrodite, and Athena. When appeal was made to Zeus, he shrewdly left the decision to the judgment of Paris.

As Paris watched his sheep on the mountain, the three aspiring Goddesses appeared to him in all their beauty. Each tempted him with the gifts in her power: Hera, wealth and authority; Athena, glory and fame in war. But when Aphrodite offered the most beautiful of mortal women to be his wife, Paris gave the prized apple to her. Unfortunately Aphrodite's choice was the lovely Helen, already the wife of Menelaus. The meeting with Helen, the seduction and flight to Troy, and the long war, at the end of which all Troy was in ashes, was the result of this human enactment of the conflict of the gods.

It was, indeed, a priest of Aphrodite from whose name the apple genus derives. Adonis, so beautiful no one could fail to love him, was the doomed offspring of an incestuous union. Aphrodite loved and tried to protect the youth, but she was unsuccessful. Devoted to hunting, he lost his life at the tusks of a mighty wild boar. When Aphrodite's priest Melus heard the news of Adonis's death, his sorrow was so great that he wanted only to die. In pity, Aphrodite turned him into an apple tree, whence the slightly altered generic name *Pyrus malus*.

As Rome took over the Greek pantheon, one of the many new deities created was Pomona, from whose name comes the art and science of all fruit culture, but particularly of apples. Statues and paintings show her carrying her pruning knife and other homely gear, and Ovid tells her charming story. She was a nymph of the woods, but, unlike her sisters, did not care to live in the forest and was always to be found in the orchard, pruning when necessary, cultivating, and bringing water to the earth when it was dry. "This was her whole delight," and she took no notice of the handsome youths who looked admiringly at her. One suitor, more persistent than the others, was Vertumnus. Fittingly, he was the deity of the changing seasons. When Pomona finally consented to be his bride, both continued to be "zealous in the care of fruitful trees."

The Romans of classical times indeed cherished apples. For centuries they had known the art of grafting and had enjoyed several apple varieties. They also used the fruit symbolically. Apples were presented as a token of love and were used in marriage rites. Horace, in the first century B.C., wrote of the custom of flicking apple seeds between the thumb and finger. If they stuck to the ceiling, it signified that one's beloved would be agreeable to courtship. "To hit with an apple" was an expression for the act of love.

Plutarch looked at the fruit in a more reserved and austere fashion. In one of the *Symposiacs,* the elegant and lofty dinner conversation focused on the variety of fruit that was at the table, and one of the company recited a verse from the *Odyssey* in which Homer singled out the apple tree particularly as "bearing fair fruit." The guests agreed that "the particular excellencies that are scattered amongst all other fruits are united in this alone."

Far less remains of the culture of the early Scandinavian peoples than of the Greek and Roman, of course, but here too the apple played a symbolic role. Archeologists have found Stone Age drawings and other relics in which apples are pictured as fertility signs.

In the pagan mythology of Valhalla and the Gods who inhabited it — such as Odin, Thor, Freya — apples had a very special function. As told in the *Edda,* the Gods were not immortal, but their life span was incomparably longer than man's. In charge of their longevity was Idun, the wife of Bragi, god of poetry, who kept a chest of apples in her care. When the Gods felt old age approaching, they appealed to her for one of the fruit, which they had only to taste to become young again.

Among the Celtic peoples, the apple was important, both in pre-Christian and later folklore. The Celtic myths are fragmentary, but their tales became intermingled with those of the people they conquered, and Christian elements mix with pagan. Connla, son of Conn, was a real chieftain of the Irish Celts of the 2nd century A.D. In legend, he is visited by a strange Goddess who urges him to come with her to Tir na m Beo, the land of eternal life. Refusing, he summons his Druid priests to use their magic against her. They are only partially successful; though the Goddess departs, she gives Connla an apple as she leaves.

From then on, Connla eats nothing except the never diminishing apple. After a month the Goddess reappears and again begs him to come with her; again he calls the Druids. This time the Goddess proclaims that the Druids will soon become powerless against a saint named Patrick. Connla seems to be instantly converted, because he leaps into the boat and speeds away with her, never to be heard from again.

In another part-Christian tale, apples symbolize the coming of a new belief. In 248 B.C. Cormac mac Art reigned as King of Ireland. One day a young man appeared before him, bearing a branch from which hung nine apples of gold. When the branch was shaken, it produced music so strange that people who heard it fell asleep. The young man told Cormac that he came from a land where there was only truth, life everlasting, and freedom from age, decay, gloom, or envy. Cormac, who coveted the apple branch, exchanged his wife, his son, and his daughter to get it. His wife and children, needless to say, were not pleased at their lord's priorities, and after some time Cormac remembered his responsibilities and set off to find them. After some adventures, in the happy way of myths, he was restored to his family, converted to the true religion, and returned to Ireland.

In the old language of the Welsh, apple translates as *aval*, thus Avalon, the Isle of Apples. In Geoffrey of Monmouth's story of King Arthur's death, Avalon appears as Insula Pomorum, where the soil is so rich that it need not be cultivated. "The isle lacks no good thing; there is no death, age, or sickness." It is ruled by three maidens, the chief of whom is Morgan le Fay. When Arthur, fatally wounded, departs from Camelot he is met by three ladies dressed in black. They carry him to Avalon in a small boat, and Arthur lives on there, safe from the ravages of death by virtue of the magical power of the apple trees.

The same Morgan le Fay, or Morgana, appears in the story of Ogier the Dane, or Holger Danske, the first king of Denmark and a paladin of Charlemagne's. His life, real and legendary, is an account of great chivalric deeds: heroism against the Saracens, loyalty to his liege lord Charlemagne, and gallantry to the ladies, all in such measure as to incur the jealousy of Morgana, who has had a proprietary interest in him from his birth. She entices Ogier to Avalon, and once he has eaten of the magic apples from the orchard, like Arthur, he is forever bound to the island to remain alive. Some versions of this have a postscript: Arthur and Ogier will be found on Camelot, alive and well, at the second coming of Christ, and the apples will return to the Garden of Eden, thus in a sense completing the mystic apple circle.

In a little-known morality tale of the Middle Ages, Peleringe (Pilgrim) and his Angel guide are surveying Heaven, Hell, and Earth. Two naked figures playing with an apple beneath a green tree are represented by the Angel as souls who have found the "right way," as symbolized by the apple. This is not the apple through which misfortune came to the world but rather the sign of Christ crucified for the redemption and restoration of the human race. The Angel explains, in quite a horticultural vein, that as the seeds of the wild apple will produce sour and bitter fruit unless a graft from a good apple tree is introduced, so does the progeny of Adam, tainted with original sin, need to be similarly dealt with. In this tale God the Father chose a suitable stock, Saint Anne, "grafted upon the fair graft of the Virgin Mary, and brought forth fine fruit." This homely metaphor must certainly have been the work of an orchardist who had himself cut many a fair graft.

Even then, some Christians took a less earnest view. A fifteenth-century English bit of verse using the apple as the symbol of lust shows another way of thinking:

> Adam lay I-bowndyn, bowndyn in a bond
> Foure thousand winter thowt he not too long
> And al was for an appil, an appil that he tok
> As priestes fynd a-written in their Book.
> Bless the time the Appil was taken!

As we come closer to modern times, the apple gains different meanings. It often appears as the all-purpose generalization, a kind of handy reference symbol used to make a point. In Langland's *Piers Plowman*, the fourteenth-century allegory, Piers gives as an example of simplicity and purity: "For God seith hit hym-self, 'Shall never good appel/Through any subtle science, on sour stock grow." The saying of the same period "Betere is appel y-yeue then y-ete" more prosaically reads, "It is better to give than to receive." And Chaucer's "The rotten apple injures its neighbors" is one of the first of many versions of this maxim.

Shakespeare, particularly, used the apple as a random symbol. "An apple cleft in two is no more twin than these two creatures," in *Henry V*. Or, "Though she's as like this as a crab's like an apple, yet I can tell what I can tell," says the Fool in *King Lear*. And in a burlesque note in *A Midsummer Night's Dream*, "[It] Somewhat doth resemble this/As much as apple doth an oyster/Or codling when it's almost an apple." And grandly, from *The Tempest:* "He will carry this island home in his pocket and give it his son for an apple."

A very expressive image that does not seem to have survived the prudishness of intervening centuries is paraphrased by Jonathan Swift:

> See how we apples swim
> While tumbling down the turgid stream . . .

This evolved from an earlier European version which even the great Martin Luther quoted, and which appears widely throughout German and Dutch writing. The best rendering is that of Sir Roger L'Estrange, who in 1692 published his *Fables:*

> Upon a fall of rain the current carried away a huge heap of apples, together with a dung hill that lay in the water-course.

> > A Ball of new-dropt Horse's Dung,
> > Mingling with the Apples in the Throng,
> > Said to the Pippin, plump and prim,
> > See, Brother, how we Apples swim!

This last line remained until the nineteenth century as a tagline identifying hypocrisy.

A story that most of us grew up with is that of the great mathematician and philosopher Sir Isaac Newton "discovering" the law of gravity. The highly colored scene is part of our baggage of mental images: Isaac Newton (not yet Sir) half reclines in an orchard; an apple falls to earth; in a Eurekalike instant, gravity is understood! Newton was a rather dreamy observer as well as a brilliant insightful scientist, but the tale is thought to have evolved from his having used the example of the apple's fall to illustrate the pull of gravity rather than as the "cause" of his insight. To-

Sir Isaac Newton (The Granger Collection)

day, in the Royal Botanical Gardens at Kew, England, there grows an apple tree, Flower of Kent, a descendant, through several grafts, of the original that was designated as "Newton's Tree." The variety is an everyday red cooking apple.

The apple has been seen for centuries as having the quality of bringing some special gift and as a means of divination. It served as a love charm until the recent past (and perhaps still does in areas remote from more pressing myth-makers such as television commercials). Seventy years ago an archeologist wrote of still current customs in the British Isles: a girl removes the peel of an apple in a long strip, throws it over her shoulder, and in its twists and turns tries to make out the initials of her sweetheart. Or, clapping seeds to her forehead, she counts the number of her family-to-be. In Montenegro, a bride attempts to throw an apple on the roof of her new house to insure that she will have a child. Perhaps the only custom that has survived modern urban life is that of bobbing for apples on Halloween, a children's game that is a harmless descendant of a long-ago Druid rite of divination.

In today's world the apple has become a ubiquitous visual symbol, often used commercially to stimulate — or cash in on — an association with domesticity and coziness, thus a rash of copyrighted apple names and designs. In language, too, the apple continues its sway. Some phrases smack of the standards of the nineteenth century, as "apple-pie order," or "all talk and no cider," a proverb Washington Irving cited to show the American people's "propensity to talk." This century, "applesauce" as a derisive comment was coined in 1921 by the late cartoonist Tad Dorgan. "Apple polisher" or "apple shiner," as a flatterer or toady, has only been in use since about 1925 and derives from the older tradition of the eager student who gives the teacher an apple. "Apple knocker" for a rustic type is also of this century; so, at the other extreme, is "Big Apple" referring to New York City, a term from jazz jargon of the 1930s.

So the eloquence of the apple persists in a world more attuned to plastic than bark, leaf, and fruit. If these rather flippant phrases lack the depth of feeling of earlier images, they are part of our world nevertheless.

Our Apples are, without Doubt, as good as those of England, *and much fairer to look to.*

P. Dudley, *Philosophical Transactions*, 1724

2. The Apple Comes to America

THE PURITAN settlers in Massachusetts, even while building the New Jerusalem, still held to memories of England as models of what home should be. Although bare survival was their first concern in the new land, the colonists also sought to re-create comforting reminders of the Old World. So they planted corn to maintain life, but apples, perhaps, to maintain hope. The growing tree made a rough cabin seem less stark, more akin to the remembered cottage of Surrey or Kent, and in a few years when the fruit found its way to the table, it became a staple of the meager diet.

Old England had cherished its apple trees even before the Dru-

ids wove spells over their branches. Indeed, in Europe the fruit
was known and used at the end of the Stone Age. As agricultural
skills developed, apples were among the first plants to become
domesticated, quite likely in the area between the Black and Cri-
mean seas. By the Roman era, a number of individual varieties
were cultivated and grafted to produce superior fruits, Pliny tells
us. But in America, the indigenous apple tree was a sour crab apple,
Pyrus coronaria, nothing like the sweet juicy fruit the colonists had
known in their native shires. Failing to find a suitable species here,
the early settlers brought seeds, and then seedlings, and even trees.

In 1625 a clergyman, William Blaxton, came to the Massachusetts Bay Colony to preach the Gospel and till the soil. On the
slope of Beacon Hill in Boston, near the corner of Beacon and
Charles streets, he laid out the first American apple orchard. Two
governors of the Bay Colony, John Winthrop and John Endecott,
also established plantings of considerable size; in 1644 Endecott
wrote plaintively of his need to replenish his stock, "My children
burnt mee at least 500 trees by setting the ground on fire neere
them."

As the colonists moved from Massachusetts into other areas,
they took their apples with them. Despite the cold of the northern
regions they found receptive soil and ideal growing conditions
throughout New England. Indeed, in parts of Maine, the newcomers would see flourishing apple trees planted by the French
who preceded them. By 1650, in Maine, Connecticut, and Rhode
Island, apple trees were growing in and near the newly established
villages.

Down in New Amsterdam Governor Peter Stuyvesant laid out
a farm he called "The Bouwerie" in the lower part of Manhattan
island. About 1647 he planted grafted trees from Holland, the
first of their kind on American soil. (The apple, unlike most
plants, will not reproduce "true to seed," and in order to insure
replication of fruit, a portion of the tree itself, usually in the form

of a bud, is introduced or grafted onto another or "stock" tree.*)

Since the native crab apple was of limited use for food, the Indians did not cultivate the tree. But the eastern tribes were skilled agriculturists, and as the colonists learned about the indigenous corn, beans, and tubers from them, the Indians, in turn, observed the foreign plants introduced by the invading English, Dutch, and French, and appropriated what they felt would serve their way of life. The apple tree, whose fruit they mostly dried, became one of the new food plants cultivated extensively by the Indians.

In 1743 a distinguished horticulturist, John Bartram, noted in his *Travels from Pensilvania to Onondago, Oswego, and the Lake Ontario* that the Indians in that western area grew apples in and around their villages. By 1790 there were several Indian orchards in the Cayuga and Seneca lakes regions. But these were in the path of the expanding white settlers, and in 1799 the U.S. Army destroyed their villages and wiped out acres of apple trees.

Further south, in the Carolina highlands, white settlers pushing beyond the seaboard regions found fine apple orchards surrounding the Creek, Cherokee, and Choctaw villages. To the west, Indians who had come in contact with the French were particularly skilled in the white man's agricultural methods; the great French missionaries taught practical arts and preached religious doctrine.

In 1863 Solon Robinson, agricultural editor of the New York *Tribune*, described the following apples, fruit of particular trees that had been developed and grown by Indians:

Tillaquah. The original tree of this magnificent fruit is still growing some four miles from Franklin, N.C. It is so great a favorite with all who pass the road, that but few remain on the tree to thoroughly ripen. Its name signifies "big fruit."

Toccoa. This apple was found in the orchard of Jeremiah Taylor, an old Revolutionary soldier living near the celebrated Toccoa

* See Chapter 6, *To Grow a Tree.*

Falls, in Habersham County, Ga. It ripens in August; is a very delicious, high-flavored fruit. Toccoa, when rendered in the English language, means "beautiful."

Cullasaga. Is a seedling from the Horse-apple, raised by Ann Bryson, who resides on the bank of the Cullasaga, or Sugartown fork of the Tennessee River, in Macon County, N.C.; a very aromatic, early winter apple. Its name signifies "sweet water," or "sugar water," and is pronounced cullasajah.

Yahoola. Was found growing on the banks of an old goldpit near Yahoola Creek, a large stream in Lumpkin County, Ga., and was brought into notice by Wm. Martin, Esqu. of Dahlonega, who informs us it is a desirable winter variety. The meaning of its name we do not know.

Chestoa. Takes its resemblance to a rabbit's head, being conical oblong in form, with one side near the calyx jutting over the other like a rabbit's nose.

Later immigrant groups planted apples too. German refugees fleeing religious persecution came to eastern Pennsylvania with fine seedlings from the Rhenish Palatinate and speedily established orchards. Lord William Penn himself was greatly interested in fruit growing and had his own orchards planted to "apple-grafts" as early as 1686.

The pattern of early white settlement in the South was, of course, very different from that of most of the other colonies. Land grants fostered the establishment of large plantations in Virginia and the Carolinas. The landowners, usually gentlemen of means with vast acres of tobacco as their money crop, also maintained large apple orchards to provide food and drink for themselves and their slaves. Some gentleman farmers, like Jefferson and Washington, dabbled in scientific agriculture and grew apples as an avocation. Our first President, whose early cherry-tree encounter is part of the folklore, did indeed prune his own trees and advocated pruning to his contemporaries as a beneficial exercise.

From the midseventeenth century on, wealthy settlers in Virginia and the neighboring colony of Maryland had orchards of imposing size and quality. In 1686 one Colonel William Fitzhugh of Westmoreland County in Virginia described his orchard of 2500 apple trees as being "of many varieties, such as Mains, Pippins, Russentens, Costards, Marigolds, Kings, Magitens, and Bachelors." Some of these, such as Pearmains, Pippins, and Russentens, we can identify as varieties that would survive several hundred years and whose characteristics we know. Others we can only wonder at.

Thus, in the South, even this early, apples were spoken of in varietal form, whereas in the North, most ordinary growers raised common seedlings. The manner in which this was done was simple. Seeds taken from the pomace — residue after apples were pressed for cider — were planted in a small protected area. From those that survived, the hardiest were then selected and transplanted to the desired place. Yet while most farmers raised their own seedling trees, even from the earliest days nurseries dealing in apples and other fruits existed for those who wanted to buy stock. In Connecticut, only a few years after its settlement, one Henry Wolcott, Jr., set up a nursery business in fruit trees. Apples were his chief offering, and from 1648 to 1653 he sold ordinary apple seedlings for two to five pence, and some few named varieties for a bit more. But it was not until the next century that nurseries became the source of most fruit trees. In 1759 two important nurseries were established, both in Long Island. One was in Oyster Bay, and the other, bearing the impressive name of the Linnaean Botanic Garden, was in Flushing.

As time passed and more settlers came, almost every homesteader had at least one apple tree, and farmers set out as many trees as they could devote land to their growing.

"Here were small farms, each having its little portion of meadow and cornfield, its orchard of gnarled and sprawling apple trees" is Washington Irving's description of farmhouses at the beginning of the nineteenth century. In an economy where "waste not, want not" was the motto from conviction as well as bitter need, the apple tree, standing perhaps on either side of the dwelling in farm or town, was utilized fruit and branch. As food, drink, forage for animals, wood for whittling, small machine parts, and finally as a lingeringly fragrant firewood, the apple tree gave full measure to whoever planted it.

The housewife made apple butter, conserves, and jellies, and she sun-dried the fruit as well. Excellent vinegar, extremely important to the Colonial kitchen, was derived from apples. As in all times, the best apples were eaten as they came. Certainly apples in cooking and baking were of great importance, giving variety and sweetness to a limited diet.

A Frenchman, Michel Guillaume Jean de Crèvecoeur, a soldier, surveyor, and explorer who settled in Orange County, New York, at the end of the eighteenth century, published his observations of rural American life in the form of letters to England.

> Perhaps you may want to know what it is we do with so many apples. It is not for cider, God knows! Situated as we are it would not quit cost to transport it even twenty miles. Many a barrel have I sold at the press for a half dollar. As soon as our hogs are done with the peaches we turn them into our apple orchards.

Describing the drying of apples, Crèvecoeur gives an endearing picture of country living.

> Our method is this: we gather the best kind. The neighboring women are invited to spend the evening at our house. A basket of apples is given to each of them, which they peel, quarter and core.

These peelings and cores are put in another basket and when the intended quantity is thus done, tea, a good supper, and the best things we have are served up. The quantity I have thus peeled is commonly twenty basket, which gives me about three of dried ones.

Next day a stage is erected either in our grass plots or any-where else where cattle can't come. Strong crotches are planted in the ground . . . poles fixed . . and a scaffold erected. The apples are thinly spread. They are soon covered with all the bees and wasps and sucking flies of the neighborhood. This acceler-ates the operation of drying. Now and then they are turned. At night they are covered with blankets . . . By this means we are enabled to have apple-pies and apple-dumplings almost the year around . . . My wife's and my supper half of the year consists of apple-pie and milk.

The other staple for the table was apple butter, and Crèvecoeur tells how this now almost unknown pungent delicacy was made and used:

We often make apple butter, and this is in the winter a most excellent food particularly where there are many children. For that purpose the best, the richest of our apples are peeled and boiled; a considerable quantity of sweet cider is mixed with it; and the whole is greatly reduced by evaporation. A due propor-tion of quinces and orange peels is added. This is afterwards preserved in earthen jars, and in our long winters is a great deli-cacy and highly esteemed by some people. It saves sugar, and answers in the hands of an economical wife more purposes than I can well describe.

The kind of apple-slicing party that Crèvecoeur describes con-tinued until the twentieth century. In many rural sections it was a "social" as traditional as the quilting bee.

After the Revolutionary War some restless Americans, alone or with their families, left the seaboard states and began to explore the trails leading westward along the natural pathways cut by the great rivers and valleys. They sought in the rich virgin lands of Ohio, Indiana, and Illinois an easier return for their labor than they could get in rock-ribbed New England. Others left communities where debts had piled up and opportunities had withered. Later they would leave in droves, but the first trickle began before the turn of the eighteenth century.

One of the earliest among this wave of pioneers was a rather extraordinary man. Unlike most, he walked alone, usually barefoot, and legend has it that though he carried barely enough to sustain himself in difficult country he had tucked away in his knapsack a packet of apple seeds from which the first of hundreds of thousands of acres of apple trees were to spring. He was, of course, John Chapman, who became known, even in his lifetime, as Johnny Appleseed, the bringer of apple trees to the Ohio River Valley.

The storybook Johnny Appleseed is a Disneyish creature gaily scattering seeds throughout the Middle West for the benefit of future settlers. The authentic John Chapman was a highly complex person whose motives can only partially be surmised. He was born in Leominster, Massachusetts, in 1774, and when he was two years old his mother died. His father had served in the Revolutionary Army but had been dishonorably discharged. He came home, remarried, and brought his two children of the first marriage to the new family, which was to grow to ten in number. But providing for his family was not the senior Chapman's strong point, and young John apparently began to support himself early. By the time he was grown, he had drifted away from Massachusetts, never to return.

The first documented sight of John Chapman in "the West" places him in the Allegheny Valley, a young man of twenty-three, barefoot and scantily clad, caught in a severe snowstorm. One story tells how he survived by making foot wrappings from his shirt and snowshoes from a pliable young beech tree. At that time he did not appear to be very remarkable to those who knew him. Like the others, he wanted land, and that was why he came walking over a hundred miles to western Pennsylvania, searching for a place to settle. According to the earliest records, after spending the winter of 1797–1798 scouting up and down the Allegheny River, he "selected a spot for his nursery — for that seemed to be his primary object."

It is hard to imagine, at this distance in time and perspective, why a twenty-three-year-old man would have wanted to set himself up in near isolation with the goal of growing apple trees. But for the pioneer, apples had even greater value than they did for the settled homesteader back East. Although their primary uses were the same — food and drink — the ubiquitous cider and apple brandy were even more important in the stressful, rugged life of the frontier. And because of their vigorous habit of growth it was feasible to plant apple trees where more difficult, less hardy fruits would not thrive. Perhaps the ease of transporting large numbers of apple seeds — opposed to, say, more cumbersome peach pits — would alone have been an important consideration. In any case, the newcomers in the western settlements chose to raise apples. The association of trees and settlers became so fused that in parts of Ohio a prerequisite for land title was the planting of fifty apple trees, or twenty peach, within three years of settling the land. The appearance of the young trees was prima-facie evidence of more than mere good intent; it indicated conquest of the land and permanence of tenure.

John Chapman planted apple trees for the remaining forty-eight years of his life. The myth is that he was the first to bring apples beyond the Alleghenies. This was far from true. Apple orchards were flourishing in western Pennsylvania, even though it was border territory, years before he appeared there. Seventy-five

Johnny Appleseed (The Granger Collection)

miles farther southwest, in Ohio, there were two existing apple orchards, one of them planted by Ebenezer Zane, whose settlement became the town of Zanesville. But what Chapman did do was move with the frontier. While others came, settled, and planted, Chapman's plantings can be traced as a kind of moving indicator on the face of the expanding country, marking the flow of Americans westward.

By the time he was thirty, his life took on the shape it was permanently to assume: he constantly crisscrossed the territory of

Ohio, Indiana, and Illinois, planting trees, selling trees, and, indeed, giving some away when his neighbors were short of cash. But if he didn't have a home, he rarely was without land. From his first try in 1797, there are on record thirty-four attempts by Chapman to lay claim to land, either leased or bought. In the case of a few of these claims he did actually have possession. In many, however, some calamity befell, either because of claim jumping by less scrupulous settlers, or because he had no money to make the necessary payments. He simply went on, always a little bit ahead of the main body of settlers, and always planting trees.

It is said no man does anything from a single motive. Chapman's activities have been variously ascribed: to some he simply appears to have had a fanciful obsession with apple planting, to others a sincere if unconventionally expressed desire for commercial success. There is still another facet, his strong and singular religious beliefs. By the time he reached his early thirties he had become a follower of Emanuel Swedenborg and a missionary of the New Church established by his rather abstruse Platonic-Christian creed. Frontier folk for whom religion signified a certain amount of emotional release and joyful hosannas found the highly intellectual, diffident style of the New Church hard going. Chapman's belief in this subtle dualism was one more element that set him apart from his fellows. But even though he failed to find many supporters, he continued to hand out tracts, often tearing apart a treasured volume to share piecemeal the word of truth with anyone who was interested. His coreligionists valued him highly, and his work was reported on in 1822 at the Fifth General Convention of the New Church.

. . . One very extraordinary missionary continues to exert for the spread of divine truth his modest and humble effort, which would put the most zealous members to blush. We now allude to Mr. John Chapman, from whom we are in the habit of hearing frequently. His temporal employment consists in preceding the settlements and sowing nurseries of fruit trees, which he avows to be pursued for the chief purpose of giving him an opportunity of spreading the doctrines throughout the western country.

Though he may not have been successful in converting his neighbors and the people among whom he passed, he was often remembered. Somehow the cloying nickname came into being, though it was at first John Appleseed, rather than the diminutive Johnny that was affixed later on. His singularity of dress and manner certainly helped. John was tall and gaunt, wore no shoes in summer and foot rags in winter, sported a hat that sometimes served as his cook pot, and preferred the woods to a bedroom, the ground to a feather bed. Among men who killed with ease, Chapman would not wittingly take the life of a mosquito or snake. Where Indians were almost universally damned, a contemporary wrote of Chapman: "He blamed the whites for all the mischief done by the Indians." And when grown men were relaxing with a jug of cider, he might be found telling a fanciful tale to a group of children gathered around him. Perhaps such nonconformity can only be absorbed in a simple community by creating a Fancy's Child. Relentlessly, the myth began to grow, and the real man's puzzling life faded from sight. The apple trees and the legends remained.

Apples were abundant, east and west. While not all the fruit was good enough for table and cooking, much of it was of top quality, given the standards of the time, and some was even good enough to export. The bulk of the apple crop, though, never went any farther than the nearest cider mill.

While we don't know what proportion of the fruit was raised for home use, for commercial sale as table fruit, and for cider, the latter, by the mid-1830s, had begun to be affected by the temperance movement. Still, the apple trees continued to bear, unheeding. Some of these were varietal apples from rich men's holdings. Robert Pell's orchard, in the Hudson River Valley of New York, had 2000 trees, mostly Pippins and Esopus Spitzenbergs, a good proportion of which got, in season, to the city markets. But this orchard was atypical in its proximity to the city, the quality of its apples, and the degree of care the trees received.

Until the appearance of the railroad, proximity to water routes or a market town pretty much dictated where the farmer could sell his produce. With the Industrial Revolution, experiments in steam railway locomotion began both here and abroad. Then, in 1831 in upstate New York, an engine and three tiny railroad cars, grandly titled the De Witt Clinton, chugged over tracks laid by the Mohawk and Hudson Line, and the era of railroading began — with profound implications for the lives of all Americans. Agriculture particularly would be changed.

Even so, it was ten or fifteen years after the De Witt Clinton's inaugural run before enough tracks were laid to affect farmers significantly. Before the railroad barons came into their own, little lines appeared throughout the East, soon proliferating westward. By 1850 the industry had begun to consolidate, and the era of rapid and cheap transportation began. By the Civil War the ten-tacles of the Road spread near and far into agricultural areas, making freight rates as much a part of the farmer's vocabulary as feed for his cattle. He had only to pack his barrels as the apples came ripe, hitch up the team, and set off for the nearest train depot.

But now, with a market, it became imperative to take a more critical look at the product. Of the carloads of apples that came to town, few could be judged "select" by present-day standards of appearance and perfection of contour and flesh. Where the modern model of excellence is a shining, large, and (in the East, almost inevitably) red fruit, in those days two factors were vital: taste and lastingness. Other times, other ways. The particularities of the different varieties of apples were recognized and the retail buyer knew which she and her family preferred. Since the whole merchandising process was more personal, and on a much smaller scale, storekeeper Jones might offer several barrels of the apples his customers found tasty, while storekeeper Brown carried other varieties. When a shipment of especially choice Esopus Spitzenbergs or Golden Pippins came in, it was a matter of some remark, and customers entering the store would be so informed. The culls — conspicuously wormy or ill-formed apples found in the barrel — were placed in a smaller bushel or peck basket and sold at a giveaway price.

The profit margin between the first-class apples and the culls was not lost on either farmer or merchant. The hitch lay in what might be called a pest-capability. A sound apple tree will bear "perfect" fruit as part of the natural reproductive process. But from the earliest stages of growth a plethora of insect and viral life forms descend on the tree, using it as a base in *their* life cycles.

Fruit growers have always been plagued by these depredations, but the extent has varied from place to place and time to time. As the country was built up, the pest population increased. In *The Farmer's Almanac* of 1848, in a report on a meeting of the Massachusetts Legislative Agricultural Group, "the Honorable John C. Gray mentioned some of the insects which attack fruit trees. The borer usually attacks the tree near the surface of the ground; the best remedy against its ravages is to examine the tree, and cut the insect out . . . The caterpillar is another troublesome insect, which, however, can be found in its nest early in the morning. The nest can be easily crushed in the hand, which is a disagreeable operation, but still is the most effectual means of destroying them. A brush with stiff wires is also very good, if used a little oftener. For the canker worm, Mr. Gray thought there was no practicable remedy which would answer for a large orchard. He had heard, however, that a gentleman, after protecting the trunks of his trees, turned his hogs into his orchard, and that they sought out and devoured the worm."

The same issue of the *Almanac* includes a suggestion by C. Newhall, Esq., of Dorchester for dealing with the curculio: "The method found most effectual for checking the ravages of this pest, is jarring them, as soon as they are noticed, by a sudden blow with a stuffed mallet, from the limbs of the tree, into a sheet held or spread beneath for the purpose of securing them, when they be crushed. About sunrise and sunset are the times recommended."

More than twenty years later, when the country had experienced a Civil War, intense industrialization, proliferation of railroads, and a rapid increase in population through immigration particularly, the *American Agriculturist* could offer its readers little more than the same hand-to-hand combat. "The annual fight must be kept up. Wherever a tent-caterpillar's nest is to be seen there is a challenge to combat. The insect will get the best of it if it is allowed time. Make it somebody's business to destroy every nest, and let that somebody be yourself if you wish it done faithfully. Jarring the trees and catching the insects upon a sheet or other convenient receptacle is the only certain mode of warfare with the curculio yet known. Begin early in the season, and early in the morning. All washes and 'invigorators' plague the inventor more than they do the curculio."

The jarring and squashing of beetles and caterpillars continued, but not indefinitely. It began to be borne in on orchardists and farmers generally that there might be other means to deal with the pests that afflicted trees and fruit.

———•———

My apple trees will never get across
And eat the cones under his pines, I tell him.
He only says, "Good fences make good neighbors."

Robert Frost, "Mending Wall"

———•———

*On the whole the industry of grow-
ing apples rests now on a more stable
and satisfactory basis than at any
other previous period.*

S. A. Beach, *The Apples
of New York*, 1903

3. The Golden Age of Apples

AT THE BEGINNING of the nineteenth century the United States,
with a population of five million, was a land of small isolated
fields and orchards. In consequence, American farmers suffered
comparatively little from crop damage due to insects and fungi.
Natural factors, such as winds, rain, sudden temperature changes,
and the birds and insects who feed on plant predators, kept a kind
of control. In addition, certain plant species were innately resistant to common pests. But as the century wore on, the economic
needs of the burgeoning population put natural constraints into
discard. As more crops, more fruit trees, more varieties were
planted together, inevitably there also appeared more pests. In-

genious, if naive, remedies cropped up, legacy of the ignorance of an earlier age. Benjamin D. Walsh, a pioneer in the field of economic entomology, wrote indignantly in 1869 of a common and highly touted practice of treating ailing apple trees by bandaging:

> The worm in fruit trees! As if fruit trees were not afflicted by hundreds of different worms, differing from each other in size, shape, color, and habits of life, time of coming to maturity, etc., as much as a horse differs from a hog. Yet the universal bandage system is warranted to kill them all. Does the apple worm bore your apples? Bandage the butt of the tree, and he perisheth forthwith. Does the caterpillar known as the red-humped prominent or the yellow-necked worm strip the leaves off? Bandage the butt of the tree and hey! presto! he quitteth his evil ways. Does the *Buprestis* borer bore into the upper part of the trunk? Still you must bandage the butt with the same universal calico, and in a twinkling he vamoseth . . . Be the disease what it will, the universal, patent, never-failing pill is certain sure to extirpate it. In obstinate cases it may be necessary to bandage the whole tree, trunk, branches, twigs, and all; but if you only apply bandages enough, the great bandage anthelmintic vermifuge is sure to be a specific against the genus worm . . . Long live King Humbug!

How truly he spoke. But if it was indeed a time of humbug, it was also the beginning of the chemical age. The entomologists had begun their work of identifying pests some thirty years before with a slow and careful description of the types of predators, their life cycles and vulnerability. They identified the two chief enemies of the apple, the apple worm (larva of the codling moth) and the fungus known as apple scab. The codling moth lays its egg, the resulting worm bores its way into the flesh of the developing fruit, and thus, the worm in the apple. The second most prevalent blight, scab, is the very descriptive term for the fungus that inhibits growth and produces patches of malformed skin on the apple.

These and other blemishes were simply taken a great deal more for granted up to the last quarter of the nineteenth century. Our standard of excellence has a different base; today's store-bought apple may be pulpy, tasteless, and soft, but it has a smooth skin, and no evidence of insect or other predation. While a perfect, blemish-free apple was certainly not unknown or unappreciated a hundred years ago, it was not the norm. But as Americans became primarily town and city dwellers, they became more finicky, as those removed from the soil tend to do. This fastidiousness was to have major consequences on marketing practices. The farmer, in order to sell his apples, would have to reckon with this new bias; to produce blemish-free fruit, panaceas more effective than bandaging and tree shaking would be required. The emerging chemical solutions were the answer.

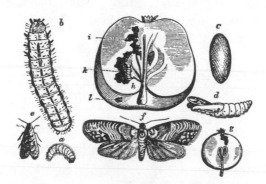

Codling Moth

For the insect predators, codling moth and others, the remedy was the arsenical poisons. The first of these was Paris Green, that old standby of early English mystery writers. In 1874 apple trees were successfully sprayed with this potent mixture. Despite its tendency to burn foliage, Paris Green did destroy insect life that preyed on the apple leaf and fruit. For the many fungi, but chiefly apple scab, there was to be another remedy. A Frenchman,

P. M. A. Millardet, who was both a grape grower and a scientist, noticed that certain of his vines were being stripped by passers-by. To discourage thievery he put a poisonous mixture of copper sulfate and hydrated lime on the vines nearest the path. He discovered that those grapes were not affected by downy mildew, although the remainder of the vineyard was devastated by the disease. Experimentation followed and in 1886 his formula was made public in the United States, where it was seized upon by apple growers particularly. It became known as Bordeaux Mixture.

Although both of the discoveries were made by Europeans, it took American knowhow to make them function so well. Instead of using the Paris Green and Bordeaux Mixture in a haphazard way with a broom, the Americans began to utilize the spray pump, thus making the essential mixture many times more effective since now the chemicals reached all the leaves. The first pump was a simple hand spray, attached to a bucket. By 1900 there were four additional models: the knapsack spray, worn on the back; a barrel pump; a gear sprayer, which had its own tank; and a power sprayer, powered by steam, gasoline, or compressed air. Thus the 1902 edition of Liberty H. Bailey's *Cyclopedia of Horticulture* could boast: "Ten years ago [spraying] was an unknown art . . . Today the American farmer leads his fellow-workers in all parts of the world in the practice of spraying." By the 1920s spraying would become accepted orchard procedure; it would be impossible to grow apples for market use unless they were sprayed, at first twice a year, and eventually, in our own time, up to fifteen times annually. Blemish-free fruit became the standard for the American consumer. Only the home grower with a few trees could have the option of deciding whether or not to spray. The commercial farmer was irrevocably tied to chemical controls.

With the chemicals to inhibit pests, the farmer now had an excellent source of revenue in his orchard. Still, most farmers, particularly on the eastern seaboard, were rarely one-crop producers and it was the exception to see 100 acres devoted solely to an orchard. In those days the dairy farmer in New York or Massachusetts might lay out a twenty-acre apple orchard on the sloping land behind his dwelling. A truck gardener or chicken-and-egg farmer could also find room for a few hundred trees — a good source of cash for little work and expense. Nursery salesmen in horse and buggy covered the back roads bearing temptingly illustrated catalogues, or, in season, bundles of dormant saplings. In 1884 "choice apple trees, two to three years old, five to seven feet" were advertised in *The Rural New Yorker* for $15.00 per hundred. In the same year fresh apples sold for $1.75 a barrel, and dried ones for 7½¢ a pound. These prices represented a considerable profit for the farmer, and cash money in hand to help pay off the ever-present mortgage. It was a rare farmer who did not avail himself of this opportunity.

The spirit of "getting ahead" was abroad in the land. The up-and-coming farmer whose father and grandfather had sworn by the quirky *Farmer's Almanac* now subscribed to the businesslike *American Agriculturist*, which provided practical day-by-day information for progressive farmers, including articles from correspondents in the South, Midwest, and West. It featured large engravings of new and tempting varieties of apples: perhaps one of the handsome Russian imports such as the Duchess of Oldenburgh, winter-hardy for the cold winds of Michigan and Maine.

Besides the journals available to the farmer, the federal government and then the individual states set up agricultural stations to spread the word of the aggressive approach to farming. Through literature and agents, they offered new ways and new products. The headquarters of the Agricultural Agency were indeed usually called experimental stations, and from them came suggestions and advice on trying various varieties, domestic and foreign. As a result, in 1900 one thousand varieties of apples were sold in the markets of the United States, according to Liberty H. Bailey, America's outstanding pomologist.

Where did all these different varieties come from, and how could they be absorbed? Part of the answer lies in the fairly specific needs of each. While it is true that apples thrive in almost any temperate region, individual varieties have a greater affinity for particular areas. Thus whole "geographical" groupings of apples were developed, and in each state the specialists sought the best for their constituency.

The New York State Agricultural Station at Geneva, New York, came into being in 1884 and began a project of planting, grafting, testing, and collecting all the examples of apple varieties it could gather. By 1903 the first volume of an extraordinary compendium was published, the result of the work at Geneva. It was called *The Apples of New York,* by S. A. Beach, and its subject was winter apples; the next year another volume dealt with early and fall varieties. Insofar as it was published by the station as part of the "Annual Report to the Department of Agriculture and the Governor," it was a New York product. But it deals with the apples of the entire Northeast and many other areas, as the names alone indicate: Virginia Greening, Kansas Keeper, Long Stem of Pennsylvania, Missouri Pippin — and even Pride of Texas.

The book addressed an immense task. It described its purpose as being "useful and stimulating to one of the most important industries of the State," but it was an encyclopedic, scientific, and even stylish presentation. For each variety named it gave synonyms, prior references in American horticultural literature, and, at minimum, a brief description. For an important variety, the description might cover several pages of quite precise details, including the growing habit of the tree itself: Of Fallawater, "*Tree* makes a moderately light root growth in the nursery. In the orchard it becomes large and vigorous. *Form* upright to roundish. *Twigs* medium in length to short, moderately stout, thick at the tips, erect; internodes medium. *Bark* smooth, bright brownish-red mingled with olive-green, finely mottled with scarf-skin; slightly pubescent. *Lenticels* moderately conspicuous, rather abundant, medium in size, usually roundish. *Buds* medium or above, roundish, obtuse, sparingly pubescent, free."

That pretty much sums up that; the earnest reader would have considerable assistance in finding out whether he had indeed a Fallawater before him, or a Buckingham ("lenticels rather scattering, below medium, generally elongated, raised").

The book also offered many historical notes: Wagener [at its best, an apple of superior excellence], "In the spring of 1791 Mr. George Wheeler brought with him from Dover, Dutchess County, N.Y., to Penn Yan, Yates County, a quantity of apple seeds which he sowed that spring in the nursery upon his farm which he was then reclaiming from the wilderness. In 1796 Abraham Wagener, from whom the name of the apple is derived, bought this seedling nursery and planted trees from it upon his place in what is now the village of Penn Yan. In 1848 it was remarked that the old tree was producing an annual and abundant yield of beautiful and delicious fruit. It continued to bear full crops till about the year 1865. After it was brought to the notice of the State Agricultural Society, the Wagener soon began to be propagated quite extensively and it has since become widely disseminated throughout the country."

There had been several books in the nineteenth century which listed and categorized apples but none with the thoroughness and authority of *Apples of New York*. From the Akin ("a beautiful dark red winter apple of pretty good quality") to the Zurdel ("hardly fair in quality and no value for the New York fruit grower") nearly 1000 apples are critically considered and meticulously described in the two volumes. Most of the varieties listed are no longer in existence, or survive in obscurity, and it is too late to know what all but a few dozen taste like. Those that have disappeared are not simply varieties of little merit; indeed, extinction seems to have come to the good, best, and inferior with a nice kind of evenhandedness.

The names represent a whole world of interests on the part of the namers. Some derive from their place of origin. Among these are Adirondack, "a promising early winter apple for Northern New York and New England"; Greenville, "suitable for general market and culinary purposes," named for an Ohio town; York Imperial, "an important apple grown commercially in the Middle Atlantic states," from York County, Pennsylvania; and Olympia, "especially desirable for culinary use," from the town of that name in Washington State.

Many more carried a person's name, either as tribute to the original grower or to the one who brought it to the attention of other growers. At least one, Milligen, "fruit of good size, rather attractive in general appearance," was named after a woman. Beach simply writes: "This variety was originated by Mrs. Milligen, near Claysville, Washington County, Pa."

The Starkey originated on the farm of Moses Starkey, in Kennebec County, Massachusetts. Ingram "Originated with Martin Ingram near Springfield, Missouri." Shannon was "A chance seedling on the farm of Wm. Shannon, Coshocton County, Ohio." And Sleight was "An apple of the Lady type which originated with Edgar Sleight, Dutchess, N.Y." All these bids for immortality, to little avail.

Many names are descriptive in part: Long Island Russet, Rhode Island Greening, Sweet and Sour, Yellow Transparent, Twenty-Ounce, Streaked Pippin, Large Early Yellow Bough. And some are dedicated to the eminent and revered: Washington Royal, Abe Lincoln, Victoria Red, Prince Albert, President Napoleon (there was an independent soul!), and Blenheim.

Some of the names are simply expressive of feeling and humor: Missing Link, Sheepsnout, Cathead, Bellyband, and Clothes-yard

Apple. In a more lyrical strain there was Mother, Autumnal Bough, Sweet Maiden's Blush, Gloria Mundi, and Beauty of the West.

Of this multitude of apples, some were clearly "better" or "best." To determine this, it is necessary to ask first what criteria are being used. There were many apples whose flesh and flavor were described by Beach and others as "best to very best": Esopus Spitzenberg, Summer Pearmain, Westfield Seek-no-further, Northern Spy, Newark Pippin, and dozens more. Yet of all the apples sold in the United States at the turn of the century, the two commercial leaders were the Ben Davis and the Baldwin. Ben Davis was the most commonly grown apple outside of the New York–New England area, where the Baldwin was primary. The most that Beach can say of Ben Davis is that "when grown in the South or Southwest, at its best it is but of second-rate quality." But always equitable, Beach adds: "The fruit is thick-skinned, does not show bruises easily, and presents a good appearance in the package after being handled and shipped in the ordinary way." Neither does he praise the Baldwin's flavor but stresses its qualities as a vigorous grower, as an apple sturdy enough to endure winter storage fairly well and to withstand the vicissitudes of shipping. The more forthright Liberty Bailey characterized the Baldwin as being "of inferior quality, and the Davis of worse." But critics notwithstanding, these two varieties continued their sway. The Baldwin was the standard apple, the moneymaker, for countless New England orchards.

The Midwest relied upon the Ben Davis. One who remembers them is Mrs. Norris Johnson, as she recollects her childhood in Benton County, Arkansas, in the 1920s. Her family's farm of eighty acres was planted three-quarters in apples, the rest in woods, pasture, and a little hay.

We lived in the center of a half mile stretch of apple trees.

The fence rows were filled with bob whites, and I always think of the fluty whistles of meadow larks and bluebirds in the spring.

The apples were largely the commercial variety of cooking apple, Ben Davis, although there were many other varieties in the orchard. Jonathan was always the favorite eating apple, crisp and juicy and flavorful. Maiden Blush ripened first but we always thought it was rather tasteless and preferred the Early Harvests. Others were Rome Beauty, Winesaps, Stark's Delicious; we had only a few young trees to try out. And a kind I have never heard of anywhere else — Florence. These trees as I remember them were not as spreading, but grew more upright. Could this have been because my father purposely pruned them that way? One tree, a greenish apple, we simply called a "sweet apple." It never produced much and I think we only ate them, or perhaps might have used some for apple sauce.

There were eight sprays in the year, beginning, I think, with a dormant spray. A team of horses drew the sprayer down the rows while men on each side sprayed the trees. The smell of the "lime sulphur" spray was fearsome and driving the sprayer as I sometimes had to do was a horrid task. The spray left a whitish-greenish surface and I remember my father coming in at night, his clothes, hat and some of his skin stained with this. I wore a deep sunbonnet to keep it off my face and felt put upon.

One year when the orchard was in full blossom, a hard freeze came and froze every apple. This was a tragedy for the county and certainly for my father who had been living on borrowed money, part of which he used to fertilize the orchard. This meant surviving another entire year with no income. That year the trees grew thick and lush, all of the growth going into the trees, and not into any fruit.

Normally during the summer, the falls were kept picked up under the trees. The worst apples were taken to the vinegar factory a few miles away. Even half rotted apples could be used in this way. Better ones were taken to the evaporator, a popular place for farm women to work and earn extra money. The apples were sliced there and dried in great racks. Children could help with this chore, picking up apples under the trees as the wagon moved slowly along, and my father was very particular that we should leave no rotting ones. He was also adamant that I should not read a book as I drove the sprayer along because I would be oblivious to the progress of the men as they sprayed and consequently delay them. Most unjust, I thought, and tiresome, but of course he was quite right.

One year, or maybe several, an old gentleman (probably much younger than I am now!) stayed with us while he pruned the orchard. I don't remember that he ever said a word, but came in and had his supper with us, then sat behind the wood stove silently until his early bedtime. He was not cross or disagreeable in any way, for I can remember that he smiled benignly at me from time to time, but evidently he just had no words. I think he must have had no home and simply lived with various people as he pruned. I have no idea what he was paid, but it surely must have been little, perhaps enough for him to live until the next winter.

At harvest time, I was in school but on coming home I used to run out into the orchard as soon as possible where the picking crews were working. Apples were brought in baskets to the "screen," a slanting slatted table at which women worked, sorting the apples — the best to go into certain baskets, the bruised or flawed ones into other baskets for a different market.

It was my chore to drive the cows up a narrow lane to the barn from their pasture, unfenced on the orchard side. My father cautioned me never to let the cows stray into the orchard as they would run under the trees knocking the apples off, or knocking down the braces that sometimes held up especially heavily laden branches. The cows seemed to take a contrary delight in breaking away and evading me to do this, while the dog and I would run frantically around to keep them in their proper bounds.

The apples we kept for home use were put in a cellar in layers of straw. It was not very deep, because Arkansas hasn't a severe climate . . . just a shingle roof, a door, a small one at one end, with the earth dug out into a rectangle. Every night my mother would fill a big bowl with eating apples and put it on the table beside which my father sat reading his newspaper — to a kerosene lamp, of course. Popcorn and apples went well together, too. We did not have a cider press but took our apples, good ones, to a neighbor. I had a horror of any wormy apples going in, after seeing what went to the vinegar factory.

For New England, Franklin County, Massachusetts, is pretty typical apple-growing country, and a man we'll call John Carr, who was born toward the end of the 1890s, is the owner of a handsome modern orchard there. He has grown apples almost all his adult life and remembers his grandfather, "who was the biggest apple grower in Franklin County . . . We grew quite a few of the varieties, but most of the apples was Baldwin."

The Baldwin was to continue its preeminence until the big freeze of 1918, when the temperature went to minus 36 degrees in the area. As Mr. Carr says, "That fixed most of them Baldwins!"

He remembers his childhood: "*Everybody* had some trees then. A lot of the old apples was originated in Massachusetts . . . Baldwin was one. Hunt's Russet, Seek-No-Further, a bunch of those old-timers. Practically all the trees anybody bought came from nurseries. They had agents around or you could write to 'em. They'd have any variety you'd want. At that time they were advocating setting 2-year-old budded trees, instead of the one-year whips we buy today. But the salesmen would come around then, and we bought 'em direct, and I expect that was one reason every farmer had his twenty or so acres in apples.

"If you was a dairy farmer you might raise corn between the rows; you was taking your land out of the dairy end of it, so for two or three years you raised some corn for feed. Then a common practice was to mow around the trees, even the young trees, and take the hay out of the orchard like it was a field. It took Baldwins or Northern Spys about ten years before they got to bearing good, and no money coming in, so you can see why farmers had to make that land pay.

"It wasn't that hard a life. We sprayed 'em maybe twice a year, arsenate, lead and sulphur, with a horse-drawn outfit then, with two hoses, and one man to drive the horses. In the old days there wasn't all these ailments that we've got today. It used to be that an apple with a codlin moth would go as a U.S. 1 apple, but now you can't have a wormy apple. Grading standards are just so different, now. Of course the ailments had to have been there, outside of the European red mite and the gypsy moth — they were imported in the late 1800s. One of these smart professors down in Harvard brought it in, to make silk I guess it was, and a pair of those moths got out. Now they're all over the East. It was one of those mistakes. But the other pests just didn't bother us too much, and we only sprayed twice a year.

"As far as picking, you didn't need any outside help, then. Your neighbors came in, those that didn't have any apple trees, and friends from town. We had a neighbor, him and his son used to pick 100 barrels a day. That's 300 bushels. You don't find that kind of help today. In the old days you packed 'em right in the orchard, brought a table, dumped 'em on the table and sorted 'em, threw out what wasn't fit, packed 'em face up in the barrel, then and there. Actually you packed a barrel bottom-side up, with the best ones at the bottom, which would be the first when it got opened, so you laid the first layer in very carefully, nice red ones.

They'd fill the barrel up a little more than full, put a press on 'em, shake a little bit and press 'em down, and that held 'em firm. Maybe a dozen or fifteen apples on the bottom would get bruised, but not too bad, and they'd be ready to move out. Then they'd get put on the way. In my grandfather's day they loaded up to Colrain City. There was a trolley there, it ran from the church to Shelburne Falls, down to the railroad tracks, to the Bridge of Flowers. They brought up freight cars, and they loaded it right there. The buyers would come around in horse and buggy, and if there was a big crop the dealers would try to knock down your price, and the growers would do the best *they* could. The dealers would buy your whole crop, and off everything would go — to Faneuil Hall Market, going very strong then, — if it was Boston. Or it might be Tredmount Co. in Providence or Norenstein from New York. A lot of 'em went for commercial cold storage, and maybe 15% of your crop, especially the Baldwin and Wealthy, went for export, England mostly. The reason why England, is that they didn't use to seem to be able to grow a decent apple. And they was willing to pay.

"In those days they was all put up in barrels — export or domestic — put up as they was picked. First of November they was all out of apples. They didn't have any cold storage on the farm, they got their money, and had the winter to do as they pleased, which is entirely different from what it is now."

Mr. Carr traces that difference to several factors: "McIntosh, Delicious, a little later Cortlands and a few other varieties, but mostly McIntosh, were coming into prominence, and Baldwins was beginning to go out. One reason was, Baldwin is an every-other-year apple, now why should you do the same work for a tree that will only give you a crop on odd years? And then there was the McIntosh. Not that it's a new apple, it's over a hundred years old, but they didn't have the spray material to grow good McIntosh till the late twenties, early thirties. And McIntosh changed a lot of things; Macs wouldn't stand barrels, they're too soft. But they bore every year, had a nice flavor, and was a good looking apple if you could keep the scab off 'em. Actually, for the customer it's looks, you got to have 'em all red, and they've got to be free from blemishes, and a fair size, not too big, not too small.

"And there was another thing. When you came to selling you'd have 25 varieties, and each grower was in the same fix. So the apple growers of New England and New York, and the Department of Agriculture people, we all got together. We met in Boston, and I happened to be one of 'em, and decided what varieties we'd promote. We decided we'd have seven varieties that was suitable for New England and New York to grow, and we'd stick to them. The customer would get used to 'em, and we'd be better off. The seven varieties were McIntosh, Red Delicious, Rome, Northern Spy, Greening, Cortland, and Baldwin."

As to why the Baldwin was retained, given its drawbacks: "We kept with the Baldwin because people still liked 'em, and we couldn't educate the public too fast. Now you've got people today who never even heard of a Baldwin."

And that was how it happened. There are people who, indeed, have never heard of a Baldwin, to say nothing of a Fameuse, Cox's Pomona, or a Yellow Transparent.

I heard a sound as of scraping tripe,
And putting apples, wondrous ripe,
Into a cider-press's gripe.

Robert Browning,
"The Pied Piper of Hamelin"

4. Cider

"THE DESIRE of the Puritan, distant from help and struggling for bare existence, to add the Pippin to his slender list of comforts, and the sour 'syder' to cheer his heart and liver, must be considered a fortunate circumstance. Perhaps he inclined to cider for the same reason that the Chaplain of Newgate, in Jonathan Wild's time, gave for his love for rum punch, — because it was nowhere spoken against in the Scriptures." Thus, John E. Russell speaking to the Massachusetts Horticulture Society in 1885.

"Incline to cider" generations of Americans did indeed. Cider was an exceptionally accommodating product: inexpensive and easy to make, long-lasting and universally accepted — young and old drinking it without question. For moderns to whom "cider" is the pleasant if rather vapid fruit drink bought usually in the fall of the year at the grocery along with the eggs and corn flakes, let it be known that this present-day phenomenon bears the same relationship to real cider as processed grape juice does to decent wine. Nor is there an appreciable difference between apple juice and present-day commercial apple cider. Both are the liquid pressed from the fruit which is then either pasteurized or treated with chemical additives for preservation and canned or bottled. Here and there in apple-growing country a roadside stand may offer the truly fresh fruit juice which it euphemistically calls cider, an artless, naturally sweet drink. Real cider — hard cider — is biteytingly, with a kind of subdued carbonation, holding flavor memories of the sour-sweet apples from which it was pressed. In the United States it is rarely sold commercially, but some who still have a press, wooden barrels, and know-how continue to provide friends and neighbors with an amiable and mildly potent liquor.

Until about fifty years ago, cider was identified as an alcoholic beverage, the natural result of fermentation. Perhaps the name is less than explicit, since most fermented fruit juices are simply called wines; cider is one exception. The sugar which is naturally in any fruit juice undergoes a process of chemical change, and the result is transformation into an alcoholic liquor. In the case of hard cider the natural alcohol content varies from 3.2 percent to 7 percent by volume — the New York State Alcoholic Beverage Commission's definition — to a possible 11 percent or 12 percent in some home farm products. Early cider, when made naturally, must have had about the same alcoholic content as today's since the varieties of apples used then are much the same as those used now, with comparable sugar content. What we don't know is how widespread was the practice of fortifying the cider with spirits of some sort. It was so often mentioned that one assumes it was indeed the usual process, in which case the potency of the final drink depended entirely on how lavish was the addition.

Worlidge's Cyder-mill of 1798 (from a contemporary drawing)

An eighteenth-century writer, John Taylor, summed up an attitude common to the time:

"Good cider would be an actual saving of wealth, by expelling foreign liquors; and of life, by expelling the use of ardent spirits. Apples are the only species of orchards at a distance from cities capable of producing sufficient profit and comfort to become of considerable object to a farmer. The apple furnishes some food for hogs, a luxury for his family in winter, and a healthy liquor for himself and his laborers all the year."

From earliest Colonial times, cider was sovereign. In the Virginia colonies the cider produced was for the use of individual plantations. Hugh Jones, chaplain to the Virginia House of Burgesses, in 1724 wrote back to England of the area's "excellent Cyder, not much inferior to that of Herfordshire, when kept to a good Age; which is rarely done, the Planters being good Companions and Guests whilst the Cyder lasts." Another Virginia observer wrote in 1785: "As to the Drink chiefly used in this collony, it is generally Cyder, every planter having an orchard and they make from one thousand to five or six thousand gallons according to their rank and fortune."

Although all the colonies enjoyed their own cider, New Jersey's product was considered the best. As early as 1682 Governor Philip Carteret wrote: "At Newark is made great quantities of cider, exceeding [in quality] any that we have from New England, Rhode Island, or Long Island." And a hundred years later a visiting Frenchman confirmed this judgment when he wrote that while Newark cider was the very best, any Jersey cider was superior to that produced elsewhere in the country. In New Jersey and other northern colonies, not all of the liquor was made for home consumption. In rural areas it might be for barter as "Half a barrel of cider for Mary's schooling," in a York village in 1805. In the cities the most common farm product mentioned in the newspapers, aside from butter and eggs, was cider.

The widespread constant use of even a low-level alcoholic liquor may seem distressing by modern standards, but it must be remembered that American attitudes were still largely fashioned by the earlier English acceptance of spirits. England had survived the gin era of the eighteenth century, but not through embracing abstinence. Although a few railed against overindulgence, use of alcohol from the cradle to the grave was endemic. In this country Dr. Benjamin Rush, a signer of the Declaration of Independence and founder of the University of Pennsylvania Medical School, made in 1792 an "Inquiry into the Effects of Spiritous Liquors upon the Human Body — to which is added A Moral and Physical Thermometer."

After describing such effects as "an universal dropsy . . . the palsy, the apoplexy and the Epilepsy" from imbibing harmful liquors, he recommends "in the room of spirits, in the first place, Cyder. This excellent liquor [is] perfectly inoffensive and wholesome."

How much was drunk? Only the wildest approximations can be made, of course, but the very fact of the barrel (usually 31½ gallons) as the common unit of measurement does indicate the extent of the consumption. In 1724 P. Dudley, writing in *Transactions*, said: "In a village near *Boston*, consisting of about 40 Families, they made near three Thousand Barrels of Cyder. This was in the year 1721. And in another town of two Hundred Families, in the same year I am credibly inform'd they make near ten Thousand Barrels." This was drunk in a year.

Some notion of amounts can be gathered from Colonel John Bellows of New Hampshire, who, from an orchard of thirty acres, made 4800 barrels of cider a year, all of which was consumed in the village nearest him. From various contemporary sources it would appear that several barrels would have been the minimum for a small farm, and possibly seven to ten barrels would not have been thought excessive for a large establishment.

If calculations of the amount of cider drunk seem Gargantuan it must be remembered that for many it represented the equivalent of *all* the diverse liquids we drink today: water, coffee, tea, fruit juice, soda, wine, beer, and hard liquor. In a poor family cider was drunk by all, at every meal. For the rich, while it may have supplemented other liquors, it had an important place both at the master's table and for his servants or slaves.

Not every one who grew apples made cider himself, but the farmer who did not took apples to a neighbor with whom he may have had a cooperative arrangement or to another farmer whom he paid in kind or in cash, usually the former. Some farmers, too, sold the juice as it was fresh-pressed, for those who wished to make their own.

As to how all this cider was made, farmers learned from their forebears, from their neighbors, and, even as now, from the printed page. When William Coxe, in 1817, wrote the first American book on fruit and its products, he began his chapter "On the Properties and Management of Cider" with these words: "This is unquestionably, the most difficult branch of the business of an Orchardist; and that on which the success of his plans must chiefly depend. It involves some principles of chymical science, not easily comprehended or explained by men of common education, yet necessary to be known to every cultivator of orchards, who aims at any degree of perfection . . ." He then goes on, having assumed the erudition of his audience, to detail the process. The properties of a cider and table apple are very different; toughness, dryness, a fibrous flesh, and astringency are all properties in a cider apple. "Apples should ripen as late as the first of November, and not later. Apples which fall fully ripe make better cider than those which are shaken . . . It is a fact generally understood, that ciders from mixed fruits are found to succeed with greater certainty than those made from one kind . . . The fruit in grinding should be reduced as nearly as possible to an uniform mass. In the operation of grinding slowly, the liquor acquires good qualities that it did not before possess."

All of this was prior to the actual process of fermentation, at which level the going does become a bit difficult:

"The fermentation of liquors has been divided into three stages; the vinous, the acetous, and the putrefactive: the first takes place only in bodies containing a considerable portion of sugar, and is always attended with the decomposition of that substance: the liquor gradually loses its sweetness, and acquires an intoxicating quality; and by distillation yields a greater, or less quantity of ardent spirit, according to the quantity of sugar and the skill of the distiller. When this fermentation proceeds too rapidly, it is sometimes confounded with the acetous; but the product of that is entirely different — when ever the fermentation, though purely vinous, becomes violent, it tends to injure the strength of the cider, by carrying off a part of the ardent spirit with the disengaged air — the acetous fermentation follows the vinous; sometimes, when the liquor is in small quantity, and exposes a large surface to the air, it will precede it — in this, the vital air is absorbed from the atmosphere, and the vegetable acid, ardent spirit, and sugar, if any remain, are alike converted into vinegar."

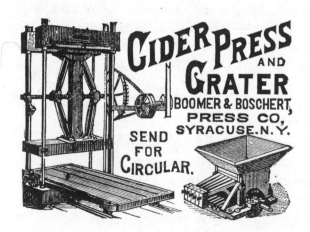

Then comes a lengthy chapter of precise instructions detailing the three processes. Exact information is given on the buildings and machinery necessary to the manufacture of cider. After such esoterica as "the pomace issues from the ullage" and "stulling with brimstone," he ends the process with a warning: "With every possible attention to the management of cider, it will require the strongest bottles to withstand its disposition to fly in our warm climate . . . [but] the breakage will seldom exceed three per cent the first summer after which there is but little risk."

Coxe's scholarly work was for many years the only American book published on fruit trees and their products, but most farmers looked to those ever-reliable publications, the almanacs, offering advice and information, cautioning moral, physical, and economic pitfalls. Each year in autumn farmers were instructed to look after their apples before heavy frost, in winter given instruction in grafting, and all year round generally kept up to the mark. In addition, *The Farmer's Almanac* of 1810 seems to have been concerned, rather atypically for the time, with standards of hygiene.

> In making cider you cannot be too clean. I always wash my mill, trough, shovels, press, hoes, and everything I use about it before I begin.
>
> And I never suffer people to bite and chew the apples and spit the pumice into the heap, nor to spit their tobacco, or blow their noses about the apples. Some men are very dirty. Nor do I suffer them to tread round the outside of the vat, without placing boards to walk upon. It is a mistaken and vulgar notion that cider will purge off the dirt and filth it receives. The best way is never to let it in.

In 1825 the editor, in the persona of Capt. Snug, is still crusading for cleanliness, and when he gives instructions for cider making, he prefaces them with another little homily:

> Pshaw, pshaw, Mr. Draggle, this is no way to make cider. I

would as soon drink from a duck pond, as to take a draught of your liquor, thus mode of rotten apples, cats dung and tobacco cuds; and your cask too smells as stenchy as a he-goat! Now the rules laid down by farmer Snug are these: "See that your mill, press, and all the materials are sweet and clean, and the straw free from must."

Cider in 1840 even became the propaganda weapon of a presidential campaign. When the dandified but capable President Martin Van Buren proposed to run again after one rocky term in office, the Whigs, party of the bankers and rising industrialists, nominated General William Henry Harrison to oppose him. Important causes were at issue — federalism, hard money, states' rights, and the foreshadowing of acute slavery conflicts — but the Whigs chose to obscure these by a nomination and campaign that directed itself to the lowest level of the electorate. Harrison was a man who stood for nothing, but he could easily and effectively be made into a symbol by the Whig king-makers. "Let him not say a single word about his principles or his creed — let him say nothing, promise nothing. Let no [one] . . . extract from him a single word about what he thinks now or will do hereafter. Let the use of pen and ink be wholly forbidden" was the advice of Nicholas Biddle, the influential power broker of the Whigs. And it was done. The very name of Harrison was ignored in favor of "Tippecanoe," the place where the army under his command years before had defeated a small band of Indians. In the campaign no principles or platform were seriously presented, only a spurious martial hero and man of the people. That he was the son of an aristocratic Virginia family, later a congressman and long-time office holder who owned a 3000-acre farm — all this was ignored. The slogan "Tippecanoe and Tyler too" (Harrison and vice-presidential nominee Tyler) was further buttressed with what came to be the central motif of the campaign: "Log cabins and cider."

The campaign against the Democrats and Van Buren attacked the inheritors of Andrew Jackson as effete eastern degenerates who drank champagne wine and daintily munched asparagus, while honest Tippecanoe ate corn pone and drank cider. And drink it down they did. A small jug of cider graced the podium of every meeting hall where Harrison spoke; free cider was passed lavishly to the crowds attending rallies, thus serving the double purpose of reinforcing the symbolism of cider as the people's drink and bestowing a kindly haze on the rhetoric to come.

Van Buren went down in a wash of cider and demagoguery, and Harrison was installed, if briefly (he died within a month), in the White House. The cider keg became for the next eighty years the cliché of cartoonists to evoke homespun virtue and down-home folks — this despite, or because of, the emergence in the 1820s of the temperance cause, that uniquely American phenomenon which was to affect the patterns of life for generations in the future. Cider, even at its most fortified, was far from a menacing liquor, but it would be swept along with the others. In addition, it was the base for a more potent drink, apple brandy, sometimes called applejack. Zadek Thompson, writing in *The Natural History and Gazeteer* of Vermont, in 1842, told the dismal story:

> In the older parts [of the state] the orchards became very extensive, the trees large, and immense quantities of apples were produced. Those were mostly manufactured into cider, in consequence of which much more cider was made than could well be consumed, in its crude state, even when it was customary for all to drink it as freely, or more so, than water, and the price abroad did not warrant the expense of transportation. Distillation was therefore resorted to, and large quantities of cider brandy were manufactured. The farmers generally having large orchards could each make, without inconvenience, from a half a barrel to two or three barrels of this liquor and when they had it in their houses, as it did not seem to have cost them much, they felt themselves at lib-

erty to use it very freely; and to this single circumstance may be traced the temporal and perhaps everlasting ruin of many of our previously thrifty farmers. This cause of ruin and misery was in the full tide of operation when the first general movement was made in New England on the subject of temperance.

The temperance movement arose from complex and contradictory currents in American life. The eighteenth-century climate of personal freedom left as a legacy a wide tolerance of personal conduct, including the right to drink what one wished. At the same time, though, there also flourished a variety of religious, social and ethical beliefs. In addition to the original Puritan strain, there was a proliferation of both indigenous and imported Christian denominations, movements, and causes. Almost all were impelled by a drive for Betterment and Higher Morality, and many had a strong aggressive component. Some church groups zealously battled for abolition of slavery, the rights of women and children, or the protection of the mentally and physically incapacitated. But in one area, all the churches, and certainly all the emerging social movements agreed: quoting Scripture selectively, they pointed an accusing finger at the evil of drink. Tracts were written and sermons were preached depicting the horrors of children abandoned and wives betrayed, and, primarily, of salvation lost.

"Temperance" itself took on a meaning far different than its literal definition; little that was temperate was done in its name. Typical of an earlier age was Benjamin Franklin's moderation. He regarded temperance as one of the thirteen cardinal virtues, and writing in the *Autobiography*, he counseled simply: "Eat not to dulness. Drink not to elevation." In regard to all moral behavior he wrote to his sister: "There are some things which I do not agree with but I do not therefore condemn them or desire to shake your Belief or Practice of them." This was not to be the attitude of the nineteenth century. It was exactly to shake the belief and practices of others that the misnamed temperance movement was directed.

By 1827 the first public resolution ever adopted in favor of total abstinence was passed by the Ontario, New York, Presbytery. (Some of the ministers, however, claimed the right to "treat their friends politely.") With this, the movement was out in the open. Its prime targets were distilled spirits, rum especially. Had not Dr. Rush provided a signpost in his "Moral and Physical Thermometer" when he described partakers of "pepper in rum" as plagued by the vices of "Idleness, Swearing, Fraud, Hatred of just government, Murder and Suicide"? But with the wave toward total abstinence the door had to be shut on the farmer and laborer's friend, the homely cider jug. So pervasive was the movement in rural areas that extremists actually called for the cutting down of trees that bore the apples from which came the drink. Many fanatics did, but not the sane-minded editor of *The Farmer's Almanac* of 1840. He exhorted the goddess of fruit:

> Pomona, I see thy red, fat cheeks with pleasure, for though I am no cider bibber, I love to feast on thy sweet codlings, thy nonpareils, and thy seek-no-farthers. Who would cut down and destroy his orchard, to keep from drinking cider? Surely the man that does this must indeed have a "weak head," a head subject to be fuddled with strange notions as well as with strong drink. Give me temperance in all things. Away with the violence of passion and folly! Verily one might as well say he will have no potatoes or rye growing upon his farm, as have no orchard; for the reason that gin and brandy, those deleterious liquids that cause so much mischief, are distilled from those materials. Nay he might even cut off his right hand, because in some incautious and evil moment it has signed his name to a bond that has caused him trouble and distress. No, no, Mr. Hasty, I shall not destroy my fine Spitzenbergs, nor my Roxbury Russetins, nor my beautiful Baldwins.

Most farmers did not go the way of the Mr. Hastys and fore-

bore to cut down their trees. Indeed, cider continued to be made, but the result of the temperance movement was that cider was no

longer the inevitable end of apple growing. It continued, although not in the quantity that was manufactured before, nor the openness. There must have been many a household where the farmer's wife, having taken The Pledge, drove her husband to either surreptitious or defiant cider making. And many other farms continued in their old accustomed ways throughout the nineteenth century. As late as 1868 the *American Agriculturist* discussed cider making with a good deal of circumspection, careful not to offend any reader.

Portable cider-mills that can be worked by hand are very convenient and useful, when there are but few cider-apples to be worked up. It often happens that a farmer has a few bushels of apples that will not keep till the time of making the main crop into cider, and in this case a portable cider-mill will enable him to use them to advantage; but when there are several hundred bushels of apples ready at one time, the old-fashioned custom of taking a load of apples and straw to the nearest cider-mill is the pleasanter, and we believe the more profitable plan. It is a kind of holiday for the boys. The apples are allowed to hang on the tree as long as the wind and frosty nights will let them. The riper they are, the better the cider. They are picked up and placed in a large heap, either in the orchard or at the cider-mill and are allowed to lie a few days to complete the ripening process, in which the starch is converted into sugar. They are then rasped or ground into pulp. If the weather is cool and the apples not quite ripe, it is better to let the pulp remain in the vat a few days before pressing out the juice. This gives the cider a higher color, makes it sweeter, and of better flavor.

Eight bushels of good apples will make a barrel of cider. The cider is usually put in barrels at once and sold while sweet.

But the *Agriculturist* was also quite willing to advise in making hard cider. They offer standard suggestions as to manufacture and storage but then come up with a highly individual treatment, "communicated by an English friend, which he assures us is attended with good results . . . Put into the barrel of cider five or six pounds of loaf sugar, and a pound of raw, lean beefsteak. Let the bung be open; keep the barrel full, so that, as fermentation takes place, the scum thrown to the surface may run through . . ." To do the *Agriculturist* justice, they comment that most readers would probably prefer to keep their intake of cider and beefsteak separate.

What of cider today? Leaving aside the seasonal apple juice masquerading as such, the likelihood of discovering real hard cider depends entirely on one's chance acquaintanceship with a backwoods cider maker; I know of none sold commercially. In the last few years a new phenomenon, apple wine, has appeared in liquor stores and in the advertising media. Its message is to the very young, and it seeks a market more accustomed to Coke than Chablis. Serious wine sellers describe it as "pop" wine, a bastardized mix of grape and fruit juice, with the dubious addition of large amounts of sugar. However, one New York City wine merchant identifies a Vermont product — based entirely on apple juice — as having the merit of an honest fruit wine.

In this regard we are behind France, which has always recognized and valued the qualities of cider as a forthright country drink. It is available as the *vin du pays* in apple-growing provinces, and high standards are maintained in its production. England, too, makes cider but it is taken less seriously; consequently it can vary widely in excellence.

Fermented cider provides the base for apple brandy, the elegant Calvados of France. In America this has been called, variously, applejack, applejohn, or Jersey Lightning, and few today know its heady charm. Commercial distilling of apple brandy, both for local and commercial use, was widespread until Prohibition. The fourteen years of that disastrous experience left many casualties, among them a real degradation of taste. At its repeal there was a rush to make alcoholic beverages, but the popular whiskies took precedence; profits were greater and less skilled workmen required than in the production of what is really America's native *eau de vie.*

Brandy is the result of several complex processes. Apple juice is expressed from the fruit, then allowed to ferment, then heated so that the liquid is turned to vapor, and then condensed. The resulting liquid is called brandy, and the art in its making is to control this process, since if distillation is continued too long, and all the base substances removed, becomes pure alcohol. Proper applejack retains enough of the fruit to remind the drinker of the simplicity of its origins even while he is imbibing the smooth and potent elixir.

Apple brandy never reestablished itself as a widely popular drink, though it is available commercially, and several companies make and market it.

In a very small way, it is also available sub rosa. In upstate New York and rural New Jersey it is said applejack may be found by those who know where to look: nothing as dramatic as Kentucky hillbillies of the 1920s, battling and outsmarting the Revenuers — just small family stills, tucked away here and there amid a plentiful supply of cider apples.

With a reviving interest in ways of the past many Americans are making, or attempting to make, their own hard cider. Though the sale of any alcoholic beverage is illegal outside the hegemony of the Internal Revenue Service, simply making it for your own use *is* allowed. So, if you grow your own apples, or have access to large quantities, and you want to reclaim some sense of adventuring into old ways, try making apple cider.

One of the handicaps is the difficulty in finding a press. A notice in the *Whole Earth Catalogue* of several years ago appealed for help in laying hands on a press. Old wooden presses have become objects of interest to antique collectors, and, hence, are prohibitive in price, and those who own an old press may prefer to display rather than use it. The strength of back and arm muscles required is a tribute to our forebears; we are, indeed, products of an effete, electric age. As yet no adequate hand press seems to have appeared on the market, and the commercial, electrically powered ones cost thousands of dollars. The best answer to the individual who has an ample source of apples and who wants to make small amounts of cider is to start with one of the fairly inexpensive, all purpose fruit and vegetable juicers, beloved of health-food devotees. (N.B. *not* the one for citrus fruits).

The best cider, or apple juice, for that matter, is made from a blend of several different varieties, each chosen for its particular attributes. In the groupings below you could select one of each for a balanced flavor.

Sweet subacid: Baldwin, Rome Beauty, Delicious, Cortland, or Grimes Golden.

Mildly acid to slightly tart: Winesap, Jonathan, Stayman, Northern Spy, York Imperial, Wealthy, R.I. Greening, Newtown Pippin. One of this type or the sweet subacid should provide the bulk of your fruit.

Aromatic: Delicious, Golden Delicious, Winter Banana, Mc-Intosh. These give fragrance and flavor.

Astringent: Florence Hibernal, Red Siberian, Transcendant, Martha. These are in the crab apple family and are hard to come by. In any case, it would represent the smallest proportion of your varieties.

The first step after assembling the ripe (but not rotten) fruit is a good washing. The type of juicer you have determines whether or not the apples need to be cut. Secure as much juice as you possibly can from the apple pulp, using whatever method is available to you. I have even known people to put apples first into a blender and then a muslin jelly bag, letting the juice drip out. This is a very slow process indeed, but they wanted to make a few quarts of cider and had no juicer. I use a home all-purpose fruit and vegetable juicer, electrically motored, and admire my 1870 wooden cider press that sits on the side porch.

So, by whatever means, you have your fresh apple juice. Pour it into spotlessly clean glass bottles or jugs, never letting it touch metal, and fill to within an inch or two of the top. Shops that cater to amateur wine makers sell special airlocks, but if you can't get these, use cotton wool plugs, placed snugly, but not so as to exclude all air. The liquid must have some air in order to ferment. Let stand at room temperature, and in about four days the sediment will fall to the bottom, while the tiny bubbles of fermentation will begin to arise. From this time on the cider may be "racked off" — separated from the sediment by siphoning.

Take a thin rubber tube about three feet long (the drugstore is a good source) and insert one end into the bottle of cider, gently sucking at the other until you taste the liquid. Once you have the process going, close your fingers over it and quickly put the now flowing tube into another, empty container, holding this below the level of the first, of course.

If you want to use your cider immediately, fine. If it's to be kept for a short time, store it in the refrigerator. There are ways of preserving cider, but I would rather not do so.

If you do not rack your cider before about ten days, it will go into another stage, frothing and overflowing. This will result in a cider with a higher alcohol content, but it requires a good bit of work and courts the disaster of exploding bottles. Frankly, it's not worth it, and I would suggest racking off in five or six days. It's a rewarding, very slightly alcoholic drink, and it's your very own.

If apple growers are to compete in the food market they have got to be able to produce efficiently . . . They need to be more cost conscious and more production efficiency minded.

C. D. Kearl, Cornell University

5. The Modern Apple: Growth of an Industry

THE DEPRESSION brought apples to the city, to the very street corners of New York, Chicago, and Boston. The vendors were the new poor who stood, usually silently, with their basket or table of shined fruit before them and the inevitable sign: APPLES, 5¢. For the one who bought, it might be a meal, and for the seller it was an honest means of surviving.

The apple growers themselves bore up rather well under the Depression. They were far along the road of agricultural industrialization, or they were out of business. Ulysses P. Hedrick, a leading orchardist and agricultural historian, wrote stoutly in

1933: "There is a flood of literature urging the industrialization of agriculture. Much of it is stuff and nonsense. From it one would glean that the object of life is to attain efficiency. Some of the happiest, most worthy, and most influential farmers are dreadfully inefficient . . . A farm is a place of peace, a place of refuge, a home, in which a calm, sane, serene life may be lived." But this was a futile harking-back to old ways, ways that had begun to change after the Civil War and would continue until Big Business and agriculture were as interconnected as the fruit and its stem. The apple growers were an organic part of the process, from planting to selling.

The Depression itself accelerated a shift in marketing methods; until the thirties, once a crop was harvested, it was sold and physically off the farmers' hands at whatever price had been negotiated. As Mr. Carr said: "In the good old days the apples were all gone by the first of November. The change began with the low prices dealers would pay at the farm. Back in the Depression years, the prices were low anyway, and the buyers couldn't sell everything they had put into cold storage, and some of them went broke. And fewer buyers showed up, so consequently you found yourself holding your apples a little longer. A few of us built air-cooled storages, and we had to find a market for those apples. Some of the buyers would still come around, but you sold 'em out of the storage, and also you'd contact the commission man in the market. Your crop wouldn't all leave the farm at one time, but would be kept, strung out for two, three months. But even then, by January you'd be just about done."

For thousands of years the availability of fresh apples had depended on the season of the year, and to some small extent, the variety. Early apples were gone by the middle of September — some grew wrinkled within a week of picking. Fall apples lost quality by November 1, and for the sturdier winter apples December 15, or perhaps Christmas, was the limit of market life. Ordinary cold storage simply meant a place where apples were kept cool — for the small farmer that place was his storage cellar; the more prosperous grower might build or adapt a special structure. Even the cavemen knew that earth kept a cooler and more constant temperature than air, generally 50° to 60° F. The simplest method of storing food so as to slow down the natural physiological and chemical processes utilized that knowledge. Apple growers constructed buildings with proper ventilation and insulation, or simply used the cellars of apple-packing houses, provided with air vents. Ice as a refrigerant was tried, but the difficulty of temperature control made it impractical.

Mechanical refrigeration was theoretically available to the grower from the early 1900s, but the deterrent was expense — it would take a 10,000-bushel production to justify a cold storage plant, and 30,000 or 40,000 bushels to make it comfortably profitable. By the 1930s, with the winnowing out of marginal farmers who grew fifteen or twenty acres of apple trees, the field was left to those with resources to maintain effective cold rooms, kept uniformly at 32°, with ample ventilation.

Shortly after World War II, a new storage technique was developed that truly made apples available all year round. This product of modern engineering is called, appropriately, Controlled Atmosphere, handily shortened to C.A. It was developed in England, as was so much of modern apple technology, and found quick acceptance here. The "control" is a process of decreasing the ratio of oxygen in the air and allowing the carbon dioxide to attain the level of 3 to 5 percent. The normal respiration of the apple is slowed down almost to a stop. Mr. Carr built his rooms in 1957, and he explains the C.A. process in layman's terms:

"Apples breathe the same as you and I do; they'd die if they didn't have air. You leave the apples in there, they use up the oxygen. Takes about two weeks to build up the amount of carbon dioxide you want. For Macs you allow the temperature to go up to 36° to 38°, but the hard apples will store better at 33°. If your room gets down below 3 percent oxygen, you open up a door and let some in. Lots of people think you add gas to the room; in fact, they used to *call* it gas storage. But you might only add nitrogen, which is just a component of air. The C.A. gives us a chance to market McIntosh the year around. Happens my wife's got a Mac in the refrigerator, just come out of the C.A., that's over nine months old, and if you was to eat it it would be a pretty good apple. And the C.A. is why the housewife can buy apples in the store, and good ones, up into the summer."

Mr. Carr's C.A. rooms are gray concrete, high and fairly forbidding looking when empty. Bristling with valves and meters, they are a far cry from a pleasing silo or a well-proportioned red barn. But beauty is as beauty does, and the grim C.A. rooms framed against the apple orchard signify more money for the grower.

Ironically, the industry seems to drag its feet publicly about the very existence of C.A. Here and there it is extolled as part of a rather timid public relations effort. Generally, however, there is no way of distinguishing apples that have undergone controlled atmosphere storage from those that have not. The city retailers may price the C.A. apple a few cents higher, but the consumer has no clue to its past. For the most part we simply expect apples to be available at all times. A sense of "season" is one of the casualties of modern living, and retailers seem loath to point out to buyers that the firm, unwrinkled fruit on their shelves was picked perhaps a half year ago and has spent the intervening months in a C.A. room.

Still another form of control is in the very size and form of the apple tree itself. For thousands of years the standard apple tree, in nature and in man's orchard, grew tall and full, perhaps as high as forty feet. Among the many kinds, however, two strains were naturally small. Some sort of diminutive apple trees were among the booty sent back to Greece by Alexander the Great in his sweep across Asia Minor. Before the Christian era, the Romans grew a dwarf apple, probably the ancestor of what became the French Paradise rootstock. As early as 1472 the literature of horticulture speaks of the Paradise apple as stock that would produce small trees. This, and soon another, Doucin, came to be known as dwarfs, a term that brings to mind trees not only small but in some way less than complete, even abnormal. It is not the case; dwarf stock can produce sturdy, abundantly bearing trees.

Since grafting was practiced from the earliest times, and since the apple tree lends itself eminently well to training, it was inevitable that horticulturists would influence the shape and size of the tree. They did this in three ways: by using a rootstock that would produce a small plant, by pruning, and by physical manipulation of the branches. In the West this art of shaping was known as "espalier," which reached its peak in the seventeenth- and eighteenth-century gardens of France's royalty. These were the elaborate artificial contrivances, often against walls, sometimes on trellises, that seemed so fitting a part of the baroque period. In this country when a wealthy landowner wished to create an aristocratic setting and imported a skilled gardener to carry out his desires, inevitably there were espaliered fruit trees. While the purpose was largely ornamental the trees did (and do) bear excellent fruit because of the amount of sun that reaches the leaves and the control of leaf and woody growth. The gardens of Williamsburg, Virginia, and similar restorations of eighteenth-century plantings demonstrate how well they blended with Georgian architecture.

In the nineteenth century the dwarf tree fell out of favor, perhaps since the formalism of the earlier age had given way to a more expansive Romanticism. When Downing, in 1845, wrote *Fruits and Fruit Trees of America* he scarcely mentioned dwarf trees and then with disparagement as being unworthy of the amplitude of American society. A lone partisan voice was Patrick Barry, of Rochester, New York, who wrote in 1851 *The Fruit Garden: A Treatise Intended to Explain and Illustrate the Physiology of Fruit Trees, the Theory and Practice of All Operations Connected with the Propagation, Transplanting, Pruning, and Training of Orchard and Garden Trees, As Standard, Dwarfs, Pyramids, Espaliers, Etc.* His book was intended to remedy the lack of knowledge of dwarf fruits, and data was presented on how economical dwarf stock would prove to commercial growers, as well as how elegant it was for the home garden. However, his advocacy (the nursery with which he was connected had the largest number of dwarf trees of the time) was to little effect; dwarf trees in this country were curiosities grown by an occasional adventuresome gardener.

The beginning of a scientific and utilitarian approach to dwarfs began in England toward the end of the nineteenth century. By 1912 the Fruit Experiment Station at Wye College in Kent had begun the work of identifying, standardizing, and propagating that would change apple growing radically. The station became known as Malling, then East Malling; the stock was derived from rooted shoots, and was known as clonal rootstock, identified with the initials M. and then E.M. and appropriate Roman numerals rather than the fanciful names of earlier times. Thus, from Siberia to Michigan, commercial growers may work with E.M. VII, XIV, XXII, or whatever their requirements, to achieve a degree of specificity impossible to imagine in prior years. Whereas formerly the name "dwarf" described everything from a four-foot bush to an almost standard tree, the work at Malling provided a base of knowledge to differentiate between the scrubby, poor-rooting specimen and the sound, early fruiting, easily managed tree that would mean greatly increased profits. Now the grower could choose from dozens of rootstocks the ones that combine best with the selected scion variety, taking into account such variables as climate, soil quality, soil nutrients, space availability, and wind patterns.

The wave of the future, say the agroeconomists, will bring smaller acreage devoted to apple growing, with greater productivity — the sprawling apple orchard, with great spreading trees, can be preserved only as displays of nostalgia, living mementos of more abundant years. In the recent past, with standard trees planted thirty to forty feet apart, the grower expected 300 to 500 bushels of apples per acre. Now he reads that in the Netherlands, something which one would be hard pressed to call a tree, known as Slender Spindelbush, yields 1500 bushels per acre of apples; 1600 of these are grown to an acre. Although only some varieties are thought suitable for this mode of planting, extension agents and others are urging consideration of all such possibilities.

There is even a process that goes beyond the usual grafting of selected rootstock and scion for still greater specificity. It is called double-working and it involves one or more additional tree parts. It includes several variations, one of which produces the Clark Dwarf, in effect a four-story tree, the result of three graft unions on a parent root. If all of this seems to have an arcane quality, it is misleading; the purpose is single-mindedly economic. Increased profits are the aim of the apple growers, just as they are for steel manufacturers or merchants; they hope for the largest net profit for the least investment of land, labor, and capital.

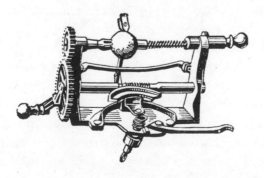

What has all this progress meant in numbers? In the chatty style of early government reports, the 1899 *U.S. Department of Agriculture Yearbook* told of progress in fruit growing that year: "In the absence of statistics on the subject, it is safe to say in a general way the year 1899 was a fairly prosperous one in fruit growing." And, "pomology not being an exact science, it is impossible to make a definitive statement of the progress" in that year. The *Yearbook* went on to note that apple growers did well, despite prevalent drought, an unusual cold wave in February, and inroads of San Jose Scale, curculio, and codling moth.

At the beginning of the 1970s, still taking into account the vagaries of the weather and the cost of the suppression of just about the same pests, statisticians were able to be more precise. In 1974, for instance, with a national production of well over six billion pounds, the commercial crop value was $520,232,000.

Where are all these apples grown? Thirty-four states have a sufficiently large production to be reported in the Department of Agriculture's Crop Reporting Board statistics. More than half of the entire six billion pounds comes from the Far West. The 1973 figures, which are fairly typical of the recent past, show 2.40 billion pounds grown in the East, .86 billion in the Central States, and 2.77 billion in the West. Washington alone produces almost one quarter of the country's commercial crop.

Western apple growing is not of course confined to Washington. Beginning up in British Columbia, and reaching down to Washington, west to Idaho and Montana, then to Oregon, California, and Utah, apples are grown as far south as Arizona. Apple orchards have existed from the days of the earliest western settlements, from seeds and trees brought by the Mission Fathers in the South and the Factors of the Hudson's Bay Company in the Northwest. In the boom days of the West with the establishment of frontier towns and cities, small-scale marketing of apples began. The transcontinental railroad was necessary before the industry could

develop into the giant it was to become. By the 1890s the Great
Northern began bringing in apple trees from the Midwest; soon
it would be shipping back carloads of fruit for city markets.

The early farm experience of New England — a few hundred
acres of diversified crops — was bypassed in the West. Apple
growing is more nearly an agribusiness in Washington than it is
elsewhere in the country, having come into existence when farm
specialization was already the trend. Rolling terrain with valleys and slopes perfect for several different varieties are found
throughout the state, and in the beginning of the twentieth century there was the usual distribution of various sorts — Ben Davis,
Gano, Wagener, Esopus, Northern Spy, Yellow Newtown — all
late-keeping winter apples and commercially valuable. The
shorter season Wealthy and Gravenstein were grown in the higher
slopes of the valleys. All this was before the ubiquitous Delicious,
the variety which won the West, and successively the South, Midwest, and even parts of the East, to become America's leading
apple.

In the East the chief apple-producing state is New York, with a
history going back to Governor Peter Stuyvesant's orchard on the
Bowerie. Although Manhattan clearly did not remain a fruit-
growing area, one apple tree the governor planted survived until
the 1860s. New York State as a whole, though, is fortunate in its
rich natural advantages, and apples can be grown almost anywhere
in the state. Commercially, the three important areas are: the
counties on either side of the Hudson River Valley — Columbia,
Greene, Ulster, Dutchess, and Orange; the area north along the
shores of Lake Champlain — Essex, Clinton, and Saratoga counties; and the western area, which includes the southern shores of
Lakes Ontario and Erie in the counties of Oswego, Wayne, Mon-
roe, Orleans, Erie, Chautauqua, and Niagara.

If eastern New York (and New England) are known as Mc-
Intosh country, in the western areas of New York the varieties

McIntosh Red

change with the marketing goals, and Macs give way to Yorks and Winesaps. Much of the Great Lakes area supplies processing, an industry which includes canned and dried apple slices, apple juice, applesauce, vinegar, apple butter, apple jelly, and a small frozen production. Of the total apple crop of the state, about two-fifths finds its way to the market as fresh fruit, the remainder being utilized in some sort of processing.

Michigan, the third largest "apple" state, devotes a slightly larger percentage to processing. Its apple growing dates back to the earliest French missionaries, who brought seeds and seedlings to establish the first orchards in the rich, well-drained lands between two great lakes. Michigan is the largest grower, by far, of the early Jonathan, bringing about half of the entire nation's crop to market. The Jonathan is a fine old apple that originated in Ulster County, New York, and was named in compliment to a gentleman of the region, Jonathan Hasbrouck. It appeared around 1820, but before long it worked its way to the Midwest, and particularly Michigan. The state is also the largest grower of McIntosh, after New York and the New England states. The remainder of its production is divided among several varieties, including Delicious, Rome Beauty, Stayman, and Golden Delicious. An important part of the apple industry in Michigan is concerned with producing nursery stock, both experimental and commercial.

Virginia, too, has a long association with apple growing. The earliest colonists in the coastal region, wishing to improve upon the crab apple they found flourishing, sent to England for scion wood and in 1622 received a good supply to graft on the wildling rootstock. The importance of the tree to the establishment of a thrifty and responsible household was underlined in 1641 when the King's representatives stipulated that the receiver of a land patent for 100 acres must plant apple trees.

Today's apple industry south of the Mason-Dixon Line is chiefly in the Shenandoah Valley, in the western part of Virginia, where the gentle slopes offer excellent growing sites. This area, like the coastal region, also has a long apple history going back, as we have seen, to George Washington and Thomas Jefferson. Virginia apples were exported in 1837 by Minister Andrew Stephenson, American representative to the English Court. When he shared them with the young Queen Victoria it is reported that she found them of such excellence that she removed the import duty on American apples, saying that "her subjects should have full access to such fine apples." In the post–Civil War years, when steamships became standard cross-Atlantic carriers, 80 percent of Virginia's apples went abroad. This market began to shrink in the 1920s, but Virginia adapted to new needs and made corresponding changes in varieties planted. The modern industry is aggressive in marketing as well, promoting an annual Apple Festival, complete with a queen and attendant hoopla.

In one particular grower, the H. F. Byrd Co., Virginia is unique. To begin with, the family firm is the largest apple producer in the country, perhaps the world. They are *the* Byrd family, whose ancestors were perhaps First among First Families of Virginia. It was a clan that included governors, royal and republican, but the family fortunes were at a comparatively low ebb by 1906 when

Harry Flood Byrd entered the apple business, buying, packing, and selling the fruit before he owned a tree. His business ability had already been demonstrated, and after buying his first 100 acres in 1912 he expanded at a formidable rate, soon including his three sons in most of his holdings and a brother in still others. Meanwhile he was able to involve himself in politics, becoming successively state senator, governor, and senator, a position he held to his death. The company's name is H. F. Byrd, Inc., and it owns some 4000 acres, on which grow over 200,000 trees in rows up to two miles long. Their apples end up as table fruit all over the East, and in processed form all over the country, as canned pie slices, applesauce and juice, and vinegar.

Whether the apple comes to market in its original packaging of skin, seeds, and stem or whether it becomes part of the huge processing industry depends on the prevailing consumer demand. For ages apples were processed, most informally, as cider or brandy, and to a certain extent as vinegar, or dried and preserved. The nineteenth century brought into being new ways of evaporating and of preserving through canning. Apple slices and applesauce became important products. In the twentieth century, baby food, frozen slices, and then apple juice, appeared commercially. Predictably, in the light of less home baking, retail apple slices for pie have shown a sharp decline — 50 percent in a recent five-year period — while apple juice consumption more than doubled from 1961 to 1971.

The last hundred years of apple growing have brought marked changes, from manipulating genes in breeding to processing fancy six packs in marketing. In every phase except weather, man is capable of controlling key factors in the industry. Along with this comes the need for decisions such as did not exist for the 1840 Capt. Snug, or even the 1940 Farmer Carr in Franklin County. The very first questions begin with tree stock.

A typical apple grower will have a mix of trees of varying ages, both standards and dwarfs. In the past, an orchard was laid out in a traditional manner; varieties appropriate to one's area were known and agreed upon. Cultural practices, such as pruning, cultivating, and feeding, were routine. And the old standard trees bore apples with no great amount of coddling, on a wide range of soil, with rather cavalier treatment except for spraying.

Today, because of the use of dwarf tree stock and the hundreds of variants possible in rootstock and grafts, the grower has to ask: How tight should I plant, and which design should I use? What varieties, which rootstock is best for me, and how?

Then, too, a new old question has arisen, a reevaluation of chemical sprays. Many commercial growers are in the "kill-'em-all" category, accepting with no misgivings an escalating use of powerful and expensive poisons, yet even they are noting the ecological disruption in the orchard itself, especially the appearance of resistant strains in the pest population. Happily, entomologists are beginning to demonstrate that other ways are possible, and many apple growers welcome this. Natural control programs, known as *integrated* biological and chemical controls, first originated in Nova Scotia in the 1940s. Now in every major apple region there are scientists at work devoted to attacking the specific pests of their area, using tactics that include sterilizing the codling moth and encouraging the ladybird beetle (*Stethorus punctum*, the predator of the destructive European red mite). Other forms of biological warfare include pitting mite against mite: as the control of the harmful McDaniel spider mite by its predator *Metaseiulus occidentalis*.

Lest hopeful, if naive, ecologists rejoice prematurely, it seems that this form of pest management must be, at present, coordinated with a controlled use of chemicals. Purists might, however, look to the work now being done in Nova Scotia. Here apples of commercial quality are being grown using biological factors exclusively in pest control, reserving chemical spraying only for the apple maggot.

In other areas a balance is sought for the precise amounts and kinds of pesticides that will not damage the "helpful" predator bug. The field is one that is ripe for exploration; it requires only some willingness to abandon our wholesale commitment to the chemical interests. It would be a comforting thought if after almost 100 years of increasing dependence on chemical poisons apple growers could begin to think in terms of the ecological relationship of the earth and the orchard, allowing the farmer to prosper and the consumer to enjoy his apple, without worm.

Where is the industry today? How does it compare with the old days, good or bad? Apples, for centuries the very synonym for fruit, are no longer first in production or use in this country. The orange — before the 1920s an exotic object that appeared in Christmas stockings or on the tables of the wealthy — has become identified as a staple American breakfast item all over the world. It takes overwhelming precedence in volume of all fruit production, fresh, frozen, canned, or reconstituted. In the first few years of the 1970s around 3 million tons of apples came to market as compared with about 14 million of citrus fruit, approximately two thirds of which were oranges. Reasons for this tremendous change in the nation's food habits and tastes are complex, but certainly a major factor is the awareness of vitamin C as a dietary necessity and the high C content of citrus fruit. Americans, whose diet patterns have been assailed by cardiologists and nutritionists, find the notion of adding the easy-to-like orange to their diet a pleasant enough way to "eat right." One could also speculate on a more subtle factor: the orange lends itself to squeezing with no great outlay of apparatus or effort, whereas apple juice is obtainable only through the utilization of a rather expensive press. Somehow the American people have become inordinately attached to drinking, whether it be hard liquor, soda, or fruit juice. (Perhaps drinking, more passively oral than eating, reflects our era?) But a major factor that has encouraged the orange growers to produce and market their huge crop is a concern for insuring against all the ills of man — from scurvy to the common cold — against which vitamin C is presumably a defense.

So numerical precedence passed to the orange, but this has at most an intangible effect on growers. In the still bountiful American market there is room for both citrus and pome fruits. The individual apple grower, like any other businessman, seeks a profitable operation, effectively balancing money, labor, and time spent against cash coming in.

Would you leave an inheritance to your children? — plant an Orchard. Would you make home pleasant — the abode of the social virtues? — plant an Orchard.

Vermont Register and
Farmer's Almanac, 1844

6. To Grow a Tree

APPLES BY THE BUSHEL, *your* apples: for the table, for pies, for pungent apple butter or crisp apple betty (thus a gain in nutrition, and money saved); a tree, handsomely setting off your house, with lovely white and pink blossoms in spring, and heavy reddening fruit at summer's end and in autumn. Strong handsome boughs in winter, a perch for cardinals and blue jays — all this from planting a tree or two? Well, perhaps.

Can you plant an apple tree? Certainly the area requirements are quite minimal and could be met by most homeowners in a small town or city suburb. Only the extreme weather zones of the United States discourage apple growing, and even in the most northerly and southerly parts of the country trees will survive, given special care. A more pertinent question is: Are you willing to support it in the style necessary to its growth? And primarily, are you willing to spray sufficiently to produce "good" fruit? That should be decided before you plant, not after.

If you have bought a house that has one or two old apple trees the chances are you will accept the fruits in a good year, make a lot of applesauce, and share the proceeds with friends. In a bad year you'll complain about the mess under the trees or simply note that *they're* not bearing this season. Since the trees undoubtedly are standards, you're not going to go to the trouble and expense of having them sprayed or of renting gear and doing the job yourself. Somehow the point of view shifts when you plant the tree yourself. The decision to spray or not to spray is a more thoughtful and deliberate one. If you're so inclined you may play Russian roulette and take the risk of disease and fungus. If you harbor an absolute ideological set against the use of chemical pesticides then you make your decision very easily. Or if you are determined to produce handsome, near perfect fruit in large quantities, and never mind the cost in ecology and dollars, there is also no need for equivocation, you will surely spray generously.

Like many of us, however, you may not fall into either of these hard and fast groups, and then you'll want to sort some things out in your mind before you decide whether or not to plant, since it must be acknowledged that spraying is essential for fruit that is of consistently good quality. As the tree develops and begins to bear, the realization comes that the odd curled leaf, the scabbed and warped fruit, and the immature droppings are not some temporary aberration that will right itself but the inevitable result of unsprayed growth. There *may* be isolated areas in the United States where the codling moth and the apple scab have not penetrated; their inhabitants are certainly to be envied. For most of us spraying is an almost foregone conclusion, the only decision being frequency.

If you decide to plant, the next choice is size. Perhaps you have in your mind's eye an imposing canopy of green, a forty-foot-high spread of bough and leaf for your children and grandchildren: then the standard apple tree in all its glory is for you. It will mature slowly, come into bearing in seven to ten years, and will establish itself with perhaps less babying than a dwarfed tree, since its far sturdier root structure tends to retain nutrients longer, particularly in dry spells. It will undoubtedly live a long and noble life, providing fruit, beauty, and shade to you and yours. If, however, your desire for an apple tree is centered on having apples, and lots of them, quickly, the dwarfed tree is indicated. Its charms for the home grower are manifold, the early return — some will bear by their third year — being only one of them. Perhaps the chief advantage for those who want only a few trees is the practicability of hand spraying. Unlike standards, which require the hiring of an elaborate and expensive rig, a few dwarf trees can be sprayed easily by one not very athletic individual using a hand spray.

Another benefit is the obvious one of space. Since self-pollination is almost unknown in apples it is foolhardy to plant only one variety of tree (unless your neighbor has thoughtfully planted one of another kind close by). And if space limitation is a factor, it should be noted that two dwarfs can be placed in little more than the area of one large tree. And even if you have a freer hand you may wish to enjoy a choice of varieties, placing, say, four or five dwarfs in the area three standards would require.

Having opted for standard or dwarf, the variety you choose will be affected by the part of the country you live in. A tree that flourishes in Michigan will do poorly in Long Island, and Maine's best may not thrive in Kansas. Your county extension agent is the one to ask, if you know him or if the office is convenient for a visit. Otherwise phone or write, requesting material on local apple varieties. You might also ask around among your neighbors as to what does well in your particular area.

In a cider-mill

The old "horse motive power" cider press, 1905

Apple trees in winter

Spraying an apple orchard in Douglas, Michigan, with a
power sprayer on a horse-drawn rig, 1915

Grading and barreling apples, 1915

A roadside farm market, 1920

Paring apples for a pie

An early twentieth-century apple-bobbing party

In the early 1930s the apple became a symbol
of the Depression

Thus armed with knowledge, you face the actual choice of variety. You have sent away for nursery catalogues *from your part of the country* and they lie before you, brilliant colors and heady prose making the senses reel. If your decision has been for standards rather than dwarfs, you will have a larger selection, but this is gradually changing. The chances are you will not be hampered by too little to choose from but rather the contrary, unless your tastes are esoteric indeed. If you want modern apples there are varieties by the score, but if you would like to try one or more of the old types a little searching in horticultural magazines will discover a nursery in your area that specializes in what one grower calls "antique apples" (see "Sources" below). Many handle both old and new, and here a caution should be noted. Some of the old apples might be interesting if you also plant other varieties, but would not be very rewarding as the bulk of one's crop. For instance, Golden Russet, which is found in many of the nurseries featuring old apples, is quite a limited, even rather dull fruit, or at least it is in our place.

The choice of old varieties must be made carefully, since for most of us there will be little chance to taste the prototypes of our selections. One way to gain more knowledge (if you find yourself in New England in the autumn months) is to take the opportunity to visit Sturbridge Village in Massachusetts, where, as an adjunct to the Colonial restoration, there is a Preservation Orchard of old varieties. Fruits of these trees are sold in the Old Country Store which is part of the Village, and used in the old cider mill. The orchard was developed by the Worcester County Horticultural Society, in consultation with the University of Massachusetts and the New York Agricultural Experiment Station. It is a project that one would hope to see replicated in other parts of the country.

Another, accidental way that you might come across old apples to taste is at a small farmstand in the countryside. In addition to the more usual kinds sold, there might be a few bags of one or two uncommon varieties. If you strike it lucky on a Labor Day week-end you might be rewarded with a bag of Gravensteins, or, later on, Spitzenbergs. When you see a small roadside stand, ask if the seller has any other varieties beside those displayed. The odd bag or two might be tucked away for local customers on the theory that city folks won't care for the lesser-known kinds. But if you can't get to decide on the basis of taste, then read the descriptions from the catalogues with care. There's no need to settle for less than "good to best" in either old or new apples.

When it comes to modern apples, let your own taste preference and the amount of cooking in your household be your guide. Start with your favorite apple "type," such as Delicious or McIntosh or Jonathan, and compare the offerings in the catalogues of reputable nurseries. You may also want to consider the "new creations" that have emerged from experimental stations and the great commercial nurseries. Many of these have rather contrived names reminiscent of Madison Avenue — Idared, Jonagold, and Spigold are some examples — but a careful reading of the catalogue will tell you the forebears, as, Spigold, cross between Northern Spy and Golden Delicious. The facts will undoubtedly be imbedded in rather florid descriptions, but don't be put off by the hucksterism. Fine new apples *are* being developed, and an inverse snobbery should not limit one's options. For instance, that old eastern standby, McIntosh, has many offspring, at least one of which, Macoun, is far superior in flavor and texture to its parent.

Another choice for the buyer is the age of the tree. Nurseries vary in their selling practices, some offering only two-year-olds, some only one-year-olds, and occasionally you may find both in the same variety. There will be a price difference, naturally, since the older tree has been cared for twelve months longer. If you have the option, and you want a bearing tree a year sooner, get the older tree.

Another factor to be taken into account is the time of flowering. Since almost no apple is self-pollinating it is assumed you will be

planting at least two trees (unless you're counting on that hypo-
thetical neighbor). If one of those flowers very early, and one
very late, you may have only a few days when cross-pollination
would be possible. The catalogue should group its selections according to summer, fall, and late varieties; if they do you can
make your own inferences. If you are unsure, your county agent
is once more your resource.

So, having weighed all these matters, you pay your money and
take your choice. Before the happy day the trees are delivered (or
picked up, if you are near a good nursery), there is work to be
done. If you have enough land to think of a miniorchard — say
a fourth or fifth of an acre — you might have eight dwarfs or four
standard trees to call your own. You can vary your crop by having
early, fall, and late apples, with perhaps some of the old favorites
and some of the newest, some for table and some for cooking. This
would yield considerably more than an individual family could
use, but in a time of new living designs it could provide a medium-
size commune or extended family with fruit for table and preserving. It could also provide youngsters or others with enough
produce for a small-size apple stand. Or if you simply want to be
very popular with your friends and relatives, a miniorchard is the
thing. If this is your wish, the Department of Agriculture's *Farmers' Bulletin 1897* (25¢) should be consulted.

For a smaller plot, and a planting of, say, two trees, the planting will perforce be near the house. The placement then will include aesthetic considerations, such as a pleasing relationship to
each other, as well as the house, but close enough — within fifty
feet — to cross-pollinate. The trees should receive full sun, and
spacing should take into account the mature foliage of years to
come. Here too the distinction between standard and dwarf must
be remembered.

If your ground slopes sharply, avoid planting at the top or the
bottom. However, if the drop is gentle, choose the higher eleva-

tion, since the important area to avoid is a "bottom" where cold air and moisture can collect. And if you are an orderly person who plans ahead, you may, after your site has been selected, enrich the earth for a season or two the way commercial growers do, by mulching and conditioning the soil.

A word on soil: For maximum commercial growth specific factors are sought, but the home owner can settle for a lot less. Perhaps one way of looking at soil requirements would be to think of avoiding extremes. While the growing tree requires water and nutrients obtainable in the soil through its roots system the apple tree is pretty adaptable. The roots go deep, so the presence of bedrock two and a half or three feet under your tree would be disastrous, as would extremely sandy or extremely heavy soil. But given average tilth, with a decent amount of water and air available, and an acidity rating of slightly acid to neutral, the apple tree will make its way.

Trees may be set out in spring or fall in warm climates, and *spring only* in cooler ones. The date will vary for the individual cold zones, but your young tree will undoubtedly be shipped with that taken into consideration. Transplanting of any living thing is traumatic, and the tree withstands the shock best if it is handled carefully. Well before the trees are delivered, "spot" the exact location, first in your mind's eye and then with a stake. You will then have a tangible "living-with" period before the trees arrive to decide if that's really where you want them. (It's something like placing the baby's crib before you give birth.)

Once you're sure of the precise places, start digging. Your aim here is not to accommodate the tree *but the root system.* The wisdom of the ages is distilled in the saying: "Don't put a fifty-cent tree in a two-bit hole." Inflation being what it is, that is multiplied considerably today, but the principle remains the same. The hole must be dug and prepared with due care and anticipation for future growth and needs. For some time all the nutrition the living plant requires will come to it from its root system. The hole must be deep and broad enough to accommodate the roots without crowding: a hole three feet across and about two feet deep will do nicely.

Presumably you have dug your holes in an already existing grass lawn if you live in the suburbs or a small town. The roots of the surrounding grass are not deep enough to compete with your young tree's nutritional needs. If, however, you've a miniorchard in a long neglected field that may have weeds or roots that go deep, they must be dealt with before the tree is planted — by rototilling, by prior spraying with a herbicide if that is your inclination, or by hand-hoeing and cultivating. *Never,* incidentally, use a herbicide near a young tree.

Take the tree out of its wrappings, cut off any dried-up or injured roots, then presoak it in a pail of water for twelve to twenty-four hours. If your soil tends to be heavy, lighten it by putting some moistened peat moss in the bottom of the hole, replace some soil, and, holding the tree with one hand so that the onionlike growth on the trunk (the graft) will be above the ground (this is vital), begin replacing the topsoil. It is important that the soil be well settled around the roots; a lack of contact will produce air pockets, which will in turn produce rot. To make sure of this, use your heel to tamp the earth, and then water slowly but well enough to reach the depth of the roots. Never apply chemical fertilizers when you plant.

Throughout the process, be concerned with protecting the tree against drying out. Try to plant it as soon as it is received, taking into account the brief soaking period. You will want to inspect it on arrival, but if it is not going to be soaked and planted immediately, rewrap it carefully. If for any reason there is a time lag of more than a few days before planting, place the trees in a shallow trench, twelve to eighteen inches deep, and mound soil around the roots. This heeling-in is simply a holding process until you are able to plant.

Postplanting concerns: If you live in a fairly rural area and small or large mammals are common, you *must* have a protective collar for your tree. The bark and emerging leaves of a young sapling are very attractive to mice, deer, and other marauders, and their depredations can be fatal to the tree. A tree trunk completely girdled by field mice is common testimony to the appetite and chewing power of this usually unseen neighbor. Baiting with poisoned grain or carrot slices might reduce the mouse population if you don't find that method distasteful, but for a few trees a simple and practical preventive for small mammal destruction is the protective collar. This is a cylinder which you can make of ¼-inch mesh hardware cloth; it should be about six inches in diameter, to enclose the growing trunk, and should be placed several inches deep in the ground and extend up to the first branch. Several styles of ready-to-buy collars are available at garden supply stores and from advertisements in horticultural magazines.

A planting note for dwarf trees: Since the root system of the very dwarf apple trees (Malling IX) is less sturdy than the standard, it is necessary to stake the tree to prevent strong winds from loosening the roots. Use a galvanized two-inch pipe or a sturdy wooden stake placed within four or five inches of the tree trunk.

A piece of cord or heavy wire covered with garden hose can be used to fasten the trunk to the stake. Make sure that this is a tie that does *not* bind as the tree grows.

As received from the nursery the spindly looking object, if it was a one-year whip, probably had no branches; if it was a two-year-old it had several side branches. It has no need of pruning, especially from those of us who are amateurs. And no pruning should be thought of for the first growing season. However, as the tree grows you should possess yourself of one or two good tools for pruning and some information. Among professionals there is great debate as to the proper, and most profitable, mode of pruning. Styles change constantly, but the home grower, not concerned with extracting the last penny from the tree, has other needs. A healthy tree will make maximum growth, not all of which should be encouraged. The various governmental agencies, state and federal, offer clear instructions on pruning, as do many gardening books. Remember that there are two purposes: (1) the elimination of dead or diseased wood and (2) the shaping of the tree to stimulate good bearing of fruit, maximum exposure to sun, and, for the home gardener, a pleasant appearance. As to when to prune, this should be done when the tree is dormant, of course, and the best time is early in spring.

Another kind of care is provided through mulching. Three ends are sought here: weed prevention, moisture conservation, and soil temperature constancy. A good thick mulch suppresses, though it does not completely prevent, weeds and grass, and it certainly is important for good tilth. However, it is one of those if-it's-worth-doing-it's worth-doing-well affairs. To be effective the mulching material — spoiled hay if you have it, grass clippings, other porous vegetable material — must be at least three to six inches deep. Do's and Don'ts for mulching include: Don't allow it to mat down and form an impervious layer; allow air and water to penetrate by loosening it occasionally. Don't put it down in one thick layer but place two or three inches down at a time.

The mulch should take care of the soil nutrients, unless you have started with less than good topsoil, in which case you might use aged manure, good rotted compost, or commercial fertilizer to feed the soil before planting. Over rich soil is not an advantage; in fact, it is to be avoided, since it will cause undue green leafy growth at the expense of fruit.

And, if success depends upon previous preparation, you're all set. The next period of time is one of watchful waiting. The first green leaves appear only weeks after planting, then in succeeding years the limbs strengthen, the trunk expands, the blossoms come, and finally there is the wonder of seeing the first fruit. If all goes well there may even be, incredibly, a problem of too *much* fruit for the individual limb to bear. In this case thinning — not of the branches but of the fruit itself — is in order. This is done judiciously where there is a cluster as you see blossom turning into fruit. Two guidelines: there should be no more than one fruit for each four to six inches of bough and the thinning should be done within twenty days of blossom time. But it will be at best several years of bearing before your embarrassment will be in the form of riches.

Fruits don't just happen, of course, in the fullness of time. Perhaps some clarification is needed as to how apples are propagated and the distinction between "varietal" and common fruit.

Reproduction in growing things can happen sexually and/or asexually. In nature apple trees reproduce other apple trees through seeds (i.e., sexually) which will carry the combined characteristics of both parents. The apple tree bears what is known as a "perfect" flower, one that has both male and female organs. However, since almost no apple will pollinate itself from its own variety — a sort of botanical incest taboo — a carrier is needed, commonly a honeybee or bumblebee. The process is as follows: the tree produces its wealth of blossoms, heavy with pollen, and the bee, attracted by the fragrance given off, goes along from tree to tree and flower to flower, leaving pollen where it touches. In these blossoms, the pollen travels down a tube to the ovary, where the sperm unites with the egg cell in the base of the flower. Seeds are then formed, surrounding tissue begins to grow, and this becomes the apple we eat.

In shape, color, and taste this apple has the characteristics of the mother tree's fruit; the seeds within it, however, are the result of the random combination of the genes of both parents. So, if you pick an apple from a McIntosh tree, you will be eating a McIntosh, but if you plant a *seed* (representing a new mix of genes) from that apple a tree will result whose fruit could conceivably be quite different. And since, as a grower, you want to control the fruit on "your" tree you must bypass the seed and plant a graft of the precise variety you desire. This is done at the nursery by taking an actual vegetative part of the mother tree and joining it onto the roots of almost any sturdy apple stock. Thus a "true" variety is assured.

Once the process of fertilization begins it usually goes along well. The hitch can be in the timing of this process. You will observe that blossoms will remain on the tree for about five days, but the discharge of sticky yellow pollen takes place for only about five hours. Usually the bees will immediately respond, and hence pollination can occur, but if the temperature is below 65° F. the bees will not "work." Commercial growers, of course, could not depend on the number of bees that would naturally occur in the given acreage of apple orchards, so they "rent" enough beehives to take care of their trees, thus helping to insure the pollination. There are other possible reasons for fruit failure, even with good pollination, such as low vigor because of past overbearing, disease, or poor nutrition.

And the final note on control of disease and fungi: The climate in your area will determine the dosage and rate of spraying. Again, your friendly county agent is the one to consult. In general, the home grower will probably want to apply an oil spray in early spring, called a dormant spray, which is a specific for scale insects and European red mite. A general-purpose spray put out by several chemical firms, detailing times and amounts, will be effective against codling moth and other pests. In late fall or early spring, a lime-sulphur spray will act as a cleanser and fungicide. But whatever you use, and however often, remember that you are handling dangerous poisons. Do not apply pesticides when they might endanger humans or animals, certainly. But also remember the beneficial insects that come to your trees, from the honeybee that is necessary to pollinate the trees to the later-season ladybug. When you spray, do it completely, making sure the whole tree is covered, but just as important, don't spray except at the appropriate times.

If despite your spraying efforts your tree begins to show patches of gray flaky substance on the bark and branches, as leaves appear, the tree has scale, which in time can be utterly destructive. The method of dealing with this pest is reminiscent of all the shaking and bandaging of the nineteenth century, but it is effective. Take an old toothbrush or stiff piece of material and simply rub off all vestiges of the scale and the eggs it protects. It works, especially if done as soon as the scale is noted.

A word of caution: let your state's Extension material be your guide rather than the chemical company's instructions on the pesticide. County agents, too, have been known to be enthusiasts in one direction or another (often prochemical), but generally state-printed material, which will give you a schedule you can follow in your area, tends to be fairly conservative.

Step-by-step, all this effort, all these decisions, may seem onerous, but the work of caring for a few apple trees is minimal. And when balanced against the manifold benefits that usually accrue it would be hard to find any good sense in *not* planting apple trees, given the small bit of earth and sun required.

And the very pace of growth, the watchful care while the young tree prepares itself for giving fruit, has some instruction for man, the taker. In a world of instant satisfactions we are pulled back into nature's rhythms. Fruits require patience: the first leaf, the first new boughs, and then, after a bit, the first apple. *That* is a moment of magic.

Sources of Old Varieties

At present the large commercial orchards seldom offer apples other than those currently popular. But here and there are indications of a movement which aims to revive, literally, old varieties. At its most accessible level this is seen in the offerings of some smaller nurseries that deal directly with the public. In addition many of the state agricultural agencies provide scions or saplings as part of their experimental work.

Nurseries

Mr. Fred L. Ashworth, Route 2, Heuvelton, New York. He sells more old varieties than any grower I have discovered, but primarily he is a plant breeder of some forty years experience, wholly immersed in grafting, acquiring, and searching for rarer apple varieties.

Baum's Nursery, R.D. 4, New Fairfield, Connecticut 06815. A considerable collection.

Bountiful Ridge Nurseries, Princess Anne, Maryland 21853

California Nurseries, Niles District, Fremont, California 94536

Grootendont Nurseries, Lakeside, Michigan 49116

Henry Leuthardt Orchards, East Moriches, New York 11940. Mr. Leuthardt's major interest is in dwarf fruit trees and particularly espalier. However, he also offers old varieties. His 25¢ catalogue is practical and full of information.

J. E. Miller Nurseries, Inc. Canandaigua, New York 14424. This is a commercial nursery that is more adventuresome than most. It offers old apple varieties in both standard and dwarf sizes, including fourteen of the latter. Free catalogue with somewhat uncritical descriptions.

Southmeadow Fruit Gardens. 2363 Tilbury Place, Birmingham, Michigan 48009. The catalogue here is $1.00.

Spring Hill Nurseries, Tipp City, Ohio 45371

Wayne Nursery, Waynesboro, Virginia 22980

Noncommercial Sources

New York State Fruit Testing Co-op Association, Inc. Geneva, New York 14456. They will propagate old varieties *on order*. This kind of custom grafting is for the very knowledgeable; you may stipulate both the rootstock and scion. Occasionally, however, they have a few odd saplings they have grafted for others left over and they will sell these. If you know what varieties you are interested in, write them, allowing a good deal of time for an answer; it will be handwritten, and neighborly. They will respond to out-of-state requests, but try your own agricultural agency as well if you are a non–New Yorker.

North American Fruit Explorers, Robert Kurle, Secretary. 87th and Madison, Hinsdale, Illinois 60521. This organization serves as a kind of clearing house for old apples, evaluating them and making information available to its members and the public.

Worcester County Horticultural Society, 30 Elm Street, Worcester, Massachusetts 01608. They are a source of information about their own work and have at times also sold scions, though not trees as far as I know.

———•—•———

But I, when I undress me
Each night, upon my knees
Will ask the Lord to bless me
With apple pie and cheese!

Eugene Field,
"Apple Pie and Cheese"

———•—•———

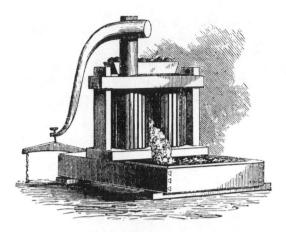

A man may think that Fameuse is a better apple than Baldwin, but if buyers want Baldwins and not Fameuse, the fruit grower must grow Baldwins.

The American Apple Orchard,
F. A. Waugh, 1908

7. A Modest Selection

THOUGH AMERICANS no longer enjoy the plenty of the early 1900s, when nearly 1000 varieties of apples actually came to market, neither are we totally without diversity. Somewhere in the United States, Gravensteins, Tompkins Kings, Northern Spys, and scores of other superb apples continue to be grown for profit and eaten for pleasure. Why then do most families tend to know and use one or two familiar kinds, the commercial leaders of their area?

Most of us limit our choices simply by the manner in which we buy, relying on the supermarket bins to supply the familiar variety. In the Northeast these will be the McIntosh, Red Delicious, and Golden Delicious, with an occasional Rhode Island Greening in

November, and from time to time some Cortlands and Rome Beautys. In the South, York, Jonathan, Delicious, and Stayman Winesaps are to be found. In the central states, Jonathan, Red Delicious, and McIntosh predominate, and in the West, again the Red and Golden Delicious, with some Jonathan and Winesap among the others offered.

Though almost all retail food buying, even in rural communities, is done in the chain stores, there is the possibility of *some* local purchasing, and consumers can exert *some* influence. When you buy apples note the grower's name and the kind you are buying, and remark on the quality. Occasionally a comparatively uncommon variety appears, new to you. Try it, and comment to the store manager. Perhaps he can ask for more of that kind or others you might want.

Mostly, though, the source of less common apples is the roadside stand, usually on the premises of an orchard and often a place where one can pick apples (for a charge) as well as buy them. Small orchards often specialize in the old sorts, and in the East and in northern California they are not difficult to track down. Typically, the Far West merchandises these matters in a superior manner: there are several apple "routes" for the tourist or local resident to explore where growers sell the rarer varieties of fruit and other apple products as well.

In buying directly from the orchard, one must usually go "in season." Apple varieties mature from mid-July to mid-November in the East and later still in the warmer areas. Growers group apples in three "seasons," summer, fall, and winter, and orchards selling retail to the consumer sell directly as the fruit is picked. In the case of summer and fall apples this is almost a necessity; early apples in particular are extremely short-lived compared with later kinds. These, however, are but a small percentage of the crop — perhaps 10 or 11 percent; most varieties grown come to maturity in late fall and winter.

The apples below are a few of the more interesting kinds still to be found, since it seems pointless to describe "the truly noble Swaar" if it is no longer available. All of these, with the exception of Macoun, are varieties which originated over 100 years ago. It is an arbitrary listing, including some of my favorites and excluding, perhaps, equally fine ones. While not every section may have all of them, perseverance should uncover two or three "new" kinds for your pleasure.

Red Astrachan

Astrachan is one of the earliest apples, typically coming into maturity by mid-July, though in northern New England they ripen by around August 20. This variety has many synonyms (including Abe Lincoln!), but most are simply variations on the spelling of its rather exotic name. It came to the United States by a circuitous route: from Russia it went to Sweden and then to England, where it first came to public attention in 1816. By 1835 the Massachusetts Horticultural Society had specimens of the Astrachan on exhibit, and it soon became one of the two or three most popular summer apples.

Astrachan is a very attractive apple of medium size with yellowish green base color overspread by shades of dark and light red, the intensity varying with the amount of sun the fruit receives. Its flesh is white, occasionally with red streaks, subacid in flavor, and aromatic. It is a particularly welcome apple for eating and cooking, appearing as it does so early. Like all summer apples, however, its charm is fleeting, within days losing savor and texture, even when chilled.

But what a delicious preview for later apples, to have a refrigerator full of bursting-fresh Astrachan, pressing to be eaten or cooked, even in the midst of summer heat.

Baldwin

Now so seldom seen, the once-abundant Baldwin paid off the mortgage of many an orchardist for over half a century, and it still kindles some loyalty in those who grow them. It is a large attractive apple, with red coloring over a yellow green base and dots of gray or white interspersed on the crimson skin. Its flesh is yellowish, and rather coarse, and with its thick skin it does not bruise easily. It bears heavily only every other year, with the intervening year being barren, or with a very scant show of fruit.

The Baldwin's early history is given in *The Horticulturist* of 1847. "This justly esteemed fruit originated in Wilmington, near Boston, in that part which now makes a portion of the new town of Somerville, in the county of Middlesex, Massachusetts. The original tree grew on the farm of a Mr. Butters, and was known for a time as the Butters apple. This tree was frequented by woodpeckers, and Mr. Butters called it the Woodpecker apple, which was soon abbreviated to the Pecker apple . . . This fruit must have been known about a century. Orchards were propagated from Mr. Butters' trees, pretty freely, about seventy-five years since, by Dr. Jabez Brown, of Wilmington, and Col. Baldwin, of Woburn, and their sons, to whom the public are principally indebted for bringing the fruit so generally into notice."

Waugh, writing in 1908, describes the success of the Baldwin: "It is exactly the apple for the ordinary man. It is an ordinary apple."

Esopus Spitzenberg

The Spitzenberg is another old apple come down in the world, though once it served as a standard of excellence for other varieties. Its origin is traced by Charles Downing to "Esopus, a famous apple district originally settled by the Low Dutch, on the Hudson, where it is still raised in its highest perfection." Throughout the nineteenth century it was held in high esteem, but today it is seldom grown in the East; even in the area of its birth it had begun to be scarce by 1900. California does grow some Spitzenbergs, though they are not shipped commercially.

It is a round, somewhat flattened apple, seldom very large, with a rich lively red skin, and irregular dots of yellowish russet. The flesh is high-flavored and very aromatic, excellent for both cooking and eating.

Its lessening popularity cannot be placed to any flaw of the fruit itself but rather to its mode of growth and the problems associated with its culture; it is a slow grower and bearer. Waugh described its tree as "poor, and subject to disease . . . Grown commercially on the Pacific Coast" in 1908.

Fameuse

The Fameuse, also known as Snow Apple, is a variety of great age which is today found almost solely in northern Vermont and across the border in the Province of Quebec. For hundreds of years the Fameuse was grown by the French in Canada and then in northern New England. It was thought by some on this continent to have originated in France and been brought here, since it was also grown in early French settlements on lakes Erie and Ontario in the seventeenth century. Conversely, European pomologists have assigned its origins to Canada, saying it was introduced in France by the early *seigneurs* who returned from exploits in the New World.

It is a handsome red apple, small, roundish and thin-skinned, its white flesh often streaked with red too. It is an eating apple almost exclusively, and comes in season in the fall.

The Fameuse, however, has a claim on history beyond its

own excellence. Unlike almost every other variety, the seed of
Fameuse tends to produce offspring quite similar to the parent;
thus, Fameuse seedlings by the thousands have been the subject
of cultivation and experimentation. What has been sought is the
Fameuse beauty and sweetness, with larger size and hardiness.
Its best-known progeny is the McIntosh, but there are many more.

Gravenstein

This is a fine apple which matures from the middle of September
until early November, and as such is included among the fall
apples. As its name indicates, it is German in origin, and there
are several versions as to its ancestry. William Kenrick, writing
in 1832, describes it as "a dessert apple supposed to have originated at Gravenstein in Holstein, and esteemed the best apple in
Germany and the Low Countries." Other accounts give it a noble
heritage, variously ascribing it to the garden of the Duke of
Augustinberg or the castle of Grafenstein.

Gravenstein is generally a medium to large apple, with a greenish yellow skin streaked with red and pink stripings. In shape it is
rather flattened — the technical term is oblate — as if pushed
down. Its flesh is whitish, juicy, tangy, yet sweet. It is excellent
for both eating and cooking.

It is found occasionally in a small orchard in the East, but in
California Gravenstein is important commercially, representing
98 percent of the summer apples grown in the Pacific Coast states.

Jonathan

This is a beautiful apple, of brilliant red color, with sometimes a
bit of contrasting yellow showing at the cavity. It is so like the

old Esopus Spitzenberg that when it was introduced in the first half of the nineteenth century it was described as "an Esopus Seedling and sometimes called the New Spitzenberg." Though Jonathan does not succeed too well in the East, it originated on the farm of a Mr. Philip Rick, of Woodstock, Ulster County, New York.

It is a winter apple, with a hardy and productive tree. Jonathan is important commercially, and while it is a staple for the Midwest, it can be found almost everywhere in the United States. Michigan is its chief grower.

Its flesh is not snowy white, but occasionally tinged with red, and it is crisp, tender, and juicy. The taste is sweet and tart, with a spicy aroma. The Jonathan is highly acceptable for both eating and cooking and deserves its long popularity.

Macoun

This is the only modern variety I know that calls forth the kind of enthusiasm that was engendered seventy years ago by a Northern Spy or a Spitzenberg. Macoun, unlike the others in this list, is the product of scientific agronomy. In 1909 a cross of McIntosh and Jersey Black (a variety I have never heard of in any other connection) was made at the New York State Experimental Orchards at Geneva. The results were many, among them one tree bearing exceptional fruit. This was named Macoun and it was presented to the world in 1923.

If its breeders sought a commercial improvement on the Mac, then this offspring would be called a failure. But in taste, aroma, and texture it is a distinctly superior apple. Large and red, it is very much like the McIntosh in appearance.

The Macoun is still largely to be found in the East; expensive New York City fruiterers usually feature the Macoun in season.

Look for it too if you are in New York's Hudson Valley in late October and November. It is not a good keeper and is seldom seen after Christmas.

But eaten out of hand, in the late autumn, the subacid yet sweet juicy Macoun could easily be the best apple many of us have a chance to bite into.

Newtown Green or Yellow Pippin

Of this now little-known apple, pomologist Charles Downing wrote in 1845: "The Newtown Pippin stands at the head of all apples, and is, when in perfection, acknowledged best in all qualities which constitute a high flavoured dessert apple." Its very name evokes apple history. The old French word *Pepin* means a pip or kernel; in English this became seed, then seedling, and then still later any apple tree which was raised directly from seed rather than from grafts and cuttings. Early in the eighteenth century on the Newtown, Long Island, estate of a Mr. Gershom Moore a chance seedling tree bore exceptionally good apples and was designated Newtown Pippin. It became so popular that the tree itself survived only till 1805 when it died from excessive cutting of scions.

It was the first apple to be exported and remained the most profitable for years. But by 1910 Waugh dismissed it as merely "a favorite in California, and in Virginia under the name Albemarle."

There are two strains of Newtown, Yellow and Green, so similar it is nearly impossible to distinguish the two. Today in California it is considered a tart cooking apple, and in New England I was sold small green rather tasteless apples as being Green Newtowns. *Sic transit gloria?* Has the variety itself deteriorated or does it not flourish in the north?

Northern Spy

The Spy evokes strong expressions from those who know it; for many it is simply *the* best apple. Larger than most, it is particularly fragrant and juicy, with flesh that is both tender and crisp. It has a bright red skin, but at the base the color is often russety.

Once the third most important apple in New York State, it is now seldom grown commercially, largely because of the uncertainty in the amount of fruit it bears. Yet for many years orchardists persisted with the Spy, knowing that even though generally a temperamental tree, when all was well the resultant fruit was superior to any.

A tree of the earliest years of the nineteenth century, its origin is traced to the farm of Oliver Chapin, near Rochester, New York, and that rather chilly temperature of its birth remains the best for its culture. It is a variety that may be found in most New England orchards which feature old apples and sometimes in the California highlands as well.

Rhode Island Greening

"The Rhode Island Greening is such an universal favourite and is so generally known that it seems almost superfluous to give a description of it," wrote Downing grandly in 1845. (But he described it nonetheless.) Today the apple is perhaps the most familiar of the old varieties, except the Jonathan, but each year fewer and fewer trees remain standing. Where it is found in the usual market sources it is presented as a pie apple, one of the best available many think. Yet 100 years ago it was considered a leading dessert apple, with culinary use secondary. This comes as rather a surprise to those who usually buy their Greenings in three-pound polyethylene bags, just the amount for one nine-inch pie. A taste while peeling and slicing puckers the tongue and raises the question: Has the apple changed, or the standard of taste? Probably neither very much, though we do generally favor sweeter fruit. The real difference, I believe, stems from the fact that Greenings now are picked deep green in color — hard and immature — undoubtedly for ease of handling. Consequently they are very tart, but when allowed to ripen on the tree they are lighter yellow in color, even with a faint blush of red, and they take on a considerably sweeter flavor, one that is juicy and rich as well.

It is a very old apple, native to America, and was well known by the beginning of the eighteenth century. The Greening originated in Rhode Island and comes by its name not only from its color but by the proximity of its progenitor to the tavern kept by a Mr. Green in the vicinity of Newport.

Roxbury Russet

Neither red, green, nor yellow, but the dull muted color from which presumably it took its name, the Russet is usually a medium to large apple, its tough skin sometimes blotchy and streaked. The flesh is yellowish and rather coarse, but juicy and often quite flavorful. But its great virtue lay, like that of a stay-at-home spinster, in its dependability. When more showy — and luscious — apples shriveled in the bottom of the bin, the Russet was still reasonably plump and smooth-skinned, still usable in spring as the only remaining fresh fruit.

To speak of "the" Russet is inexact; there have been dozens of subvarieties. The best-known of the existing ones seems to be the Roxbury, or Boston, Russet, which originated in Massachusetts in the seventeenth century. Another Russet, better-tasting though commercially unsuccessful, is the Golden Russet (which Downing says was also known by "the uncouth name of Sheep-nose"). In England its roots are so ancient that along with Pippins and Costards, Russetins were just about synonymous with all apples.

Today Russets are not grown commercially to any extent, but in the New England orchards that specialize in old varieties both Roxbury and Golden Russets are to be found.

Tompkins King

Fittingly for an apple first named King, this is a very large handsome fruit with a prevailing red color that barely covers the underlying yellow. Like the Baldwin, it is a spotted apple, with numerous white and russet dots.

Around 1800 in Warren County, New Jersey, a fruit grower named Jacob Wyckoff discovered a superior seedling tree which he called King and subsequently brought to Tompkins County, New York. It was given its full name to distinguish it from other King apples. Even at its peak it was never a widely grown apple, being susceptible to both winter kill and sun scald. Where it did succeed, though, it was always among the group of "fancy" apples bringing a premium price.

It is crisp and tender, with a juicy aromatic yellowish flesh. Small orchards in New England and New York still occasionally grow the Tompkins King.

Wealthy

Little known in the country generally, the Wealthy was the product of a search for an apple that would grow in the cold and wind of the Northwest prairie region. Around 1855 Peter M. Gideon of Excelsior, Minnesota, began a systematic planting program, producing thousands of trees each year. Each winter the cold killed them off, till at the end of ten years only one remained, from the seed of a crab apple he had obtained from Maine. The resulting seedling Gideon named Wealthy, and it was the forebear of the apple that succeeded so well in the frigid north central states.

It is a beautiful apple, quite large, striped and splashed with red. The flesh is whitish, crisp, very juicy, and agreeably tart. I have tasted it in New England, where it is to be found in some of the small orchards, and it is indeed a pleasant, if not exciting, apple. It comes in season in October and is not a good keeper in ordinary storage.

York Imperial

The York is now primarily found in the mid-Atlantic and southern states, where it is a commercially important apple, though in earlier years it was grown further west as well. As a child in Washington, D.C., I remember it was a common apple, along with the sprightly Winesap. Its origin is traced to the early part of the nineteenth century, on a farm in York County, Pennsylvania. It began life as Johnson's Fine Winter, and it was the great pomologist Charles Downing who christened it Imperial in what was surely an act of puffery.

The color is greenish yellow but so overcast with a red blush that it appears to be a red apple from a distance. In the cavity can be seen grayish or mottled russet streaks. The flesh is mildly sweet, aromatic, and juicy.

These fourteen varieties simply represent apples I find interesting, but they do not include the three leading varieties in the

United States today. These are: Red Delicious, Golden Delicious, and McIntosh.

Despite the similarity of name, Red and Golden Delicious are in no way alike, but both epitomize a changing public taste in favor of sweetness. Red Delicious is a most distinctive-looking apple, long, almost oblong-conic in shape, with vertical ridges. It was discovered in 1881 by Jesse Hiatt in Peru, Iowa, and growing rights were sold to the giant Missouri nursery, Stark Brothers, who named and merchandised it. In 1915 and in 1921 came two mutations, both deeper in color.

At its best Delicious is crisp, sweet, and aromatic. Those who buy it, and it accounts for about one out of four apples sold, almost always consider it an "eating" apple, though it can also be used in a salad where a sweet flavor is wanted. In any event it is of no use in cooking, disintegrating into a bland pulp.

The Golden Delicious, a chance seedling discovered in 1900, was also sold to Stark Brothers. It is large and handsome, with pale yellow or golden skin, the only nonred apple to succeed in the popular market. It dresses up a fruit bowl, giving nice contrast to other apples. As to flavor I find it very hard to say anything positive about the Golden Delicious. Perhaps in the West where it is chiefly grown, fresh from the tree, it is less mealy and more characterful than the ones I am accustomed to. It is the second largest selling apple in the United States.

The McIntosh, third most widely sold nationally, is the chief apple of the East. Its beginnings go back to Ontario, Canada, to the homestead of John McIntosh in the year 1810. An exceptional tree was discovered on the land, a tree whose apples became so popular that grafts of it soon spread to New York State. By 1870 McIntosh's son had named the variety McIntosh Red, and was selling it in his commercial nursery. Good as the apple was it required the chemical age to enable it to survive unscabbed and worm-free. When that was possible, by the first decade of the new century, New York and New England growers were able to present to the market an attractive, sweet, and generally dependable fruit.

McIntosh is a better apple than the Baldwin it had begun to replace by around 1920, and it was very warmly received by the public. It has aroma, beauty of color, an attractive roundish shape, and is generally of medium to large size. Though thought of as a sweet apple it has a nice subacid bite and good texture when fresh. Its drawbacks to my way of thinking are its toughish skin and poor keeping quality. This is not too serious a drawback, though; when McIntosh is past its prime for eating it makes an excellent cooking apple, either in combination or alone.

A word on a few other varieties: The Winesap is a fine old apple, red, tart, and juicy, but not large enough to prosper commercially. Stayman Winesap has its good qualities and is slightly larger. Both grow best in climates warmer than New England. Grimes Golden is primarily a Midwest dessert apple of very good quality, large, handsome, and yellow. The Cortland is similar to McIntosh though not as sprightly or juicy. The Lady apple is a

Grimes Golden

beautiful miniature fruit used for dessert or decoration. Yet it is a fine old apple, crisp and juicy, which has been cultivated in France for hundreds of years as the Api. Its base color is a creamy white or yellow splashed with red.

At the other extreme in size is the commercially successful Rome Beauty. It is the largest apple sold, with thick, almost totally red skin. One would not choose it for eating raw, but Rome has the great virtue of not disintegrating in cooking. This, combined with its size, has made it the most commonly used baking apple.

A foreign apple that has begun to appear regularly in big city markets, at least, is the Granny Smith, yellowish green, tart, and juicy. It arrives — from Australia — in the late spring, when any domestic apples available are from Controlled Atmosphere bins, and it remains through the summer. An excellent eating apple, though usually found at premium prices.

A Guide to Buying

Variety	Description and Use
Wealthy	tart, suitable for cooking, good for eating if fully ripe
Jonathan	all-purpose, including eating
Delicious	dessert, salads
Grimes Golden	excellent, dessert and cooking
McIntosh	juicy, sweet, tends to disintegrate in cooking
Cortland	all purpose, not strong in character
Golden Delicious	dessert
R.I. Greening	tart, all cooking, best for pies
Stayman Winesap	all purpose, spicy

York	keeps its shape, good for pies and all cooking where this is important
Baldwin	all-purpose
Rome Beauty	bland, but satisfactory for baked apples
Northern Spy	superior for all purposes

Bibliography

In this chapter the following authorities were consulted for historical background:

Barry, Patrick. *The Fruit Garden:* New York, 1851.

Beach, S. A. *The Apples of New York:* Albany, N.Y., 1905.

Downing, A. J. and Charles. *The Fruits and Fruit Trees of America,* 1847.

The Horticulturalist and Journal of Rural Art and Rural Taste, Albany, N.Y., 1847.

Kenrick, William. *The New American Orchardist,* 1832.

Waugh, F. A. *The American Apple Orchard,* 1908.

All human history attests
That happiness for man — the hungry sinner! —
Since Eve ate apples, much depends on dinner.

Byron, "Don Juan"

8. Apples Are for Eating

WHEN ADAM DELVED and Eve span, they and the children ate
what was available, when they were hungry. The elevating notion of eating "what was good for you" arose presumably when
a surplus of edibles allowed people the option of deciding that
certain foods were beneficial, while others were simply palatable.
The judgment was based on instinct: what made a person *feel*
good was decisive, and in that light apples were always highly
regarded. The maxim about keeping the doctor away has a long
ancestry going back at least to the late medieval version: "Eat
an apple 'fore going to bed, Make the doctor beg his bread."

But typically it was the nineteenth century, with its concern for sluggish livers and lazy bowels, and little awareness of food chemistry, that cherished the apple as particularly conducive to good health. Thus, in 1881 *The Household,* a monthly published in Brattleboro, Vermont, advised its readers:

"Let every family, in autumn, lay in from two to ten or more barrels, and it will be to them the most economical investment in the whole range of culinary supplies. A raw, mellow apple is digested in an hour and a half, whilst boiled cabbage requires five hours. The most healthful dessert that can be placed on the table is baked apple. If taken freely at breakfast, with coarse bread and butter, without meat or flesh of any kind, it has an admirable effect on the general system, often removing constipation, correcting acidities, and cooling off febrile conditions more effectively than the most approved medicines. If families could be induced to substitute the apple for the pies, cakes, candies with which children are too often stuffed, there would be a diminution of doctors' bills sufficient in a single year to lay up a stock of this delicious fruit for a season's use."

The writer continues confidently: "The acids of apple are of signal use for men of sedentary habits, whose livers are sluggish in action, those acids serving to eliminate from the body noxious matters, which if retained, would make the brain heavy and dull, or bring about jaundice or skin eruptions and other allied troubles. Some such experience must have led to our custom of taking apple sauce with roast pork, rich goose, and like dishes. The malic acid of ripe apples, whether raw or cooked, will neutralize any excess of chalky matter engendered by eating too much meat."

Fifty years later, in 1923, *The Rural New Yorker* informs us that the apple "substitutes in every requirement of life from pill to polisher . . . Where can you find a more agreeable form of necessary acids and iron than may be found in a good apple? Everyone of normal mind will understand the value of the apple as a pill, but not all know it as a polisher. It will put bloom and beauty on the skin! . . . And as a combined toothpaste and brush the apple has few equals . . . The act of eating a mellow sour apple will brush the teeth and put the acid where it is most needed; eating two good, mellow apples each day will prove the most useful exercise you can take up."

Eating even one good mellow apple, biting into the crisp flesh, catching the wayward drops of nectar with the tongue — if not useful exercise is clearly good fare. But there are times when other ways, almost as simple, suit better. Here are some recipes:

Apples in the Raw

Have these as an accompaniment to drinks, especially in the fall when the first of the new apple crop is in. For want of a simpler word they can be called

APPLE CANAPES

1. Alternate ¾″ slices of unpeeled tart apples with equal sized slices of cheese, an aged Vermont cheddar for a New England touch or a milder Gouda or Edam. Keep the cheese at room temperature for about an hour before serving, and the apples cold.

2. Core as many apples as needed but do not peel. Slice horizontally about ½″ to ¾″ thick. Soften seasoned cream cheese with a little milk or cream and spread it on the slices. Top with a very thin sliver of sweet onion.

3. 1 cup cottage cheese
    1 rounded teaspoon celery seeds
    1 garlic press squeeze of onion juice
    3 to 4 apples, cored and sliced horizontally

Mix cheese and seasonings and spread on unpeeled apple slices.

4. 1 cup of cottage cheese, cream cheese or half-and-half
    ¼ cup celery, finely cut
    1 teaspoon onion, finely cut
    1 teaspoon curry powder
    salt and pepper to taste
    3 to 4 apples, cored and sliced

Mix cheese and seasonings well and spread on slices.

5. To the above recipe add ¼ cup blue cheese and omit curry.

As you can see, the permutations are many — and you can add your own. For instance, almost any dip with a cheese base, if made with a little less liquid, can serve on apple slices.

Raw apples combine well in salads. The best known, and the butt of sophisticated gastronomic jibes, is Waldorf Salad. In its defense it should be said that its purpose is not that of a green salad — a foil to a main course at dinner — but quite otherwise. With a bowl of soup and a piece of cheese it makes a nice lunch or light supper. There are times, after all, when the tearoom meal has its comforting aspects. Take it or leave it, here is

WALDORF SALAD
for 2

> 2 medium red apples, unpeeled
> 2 celery stalks, cut up
> ¼ cup walnut meats, cut up
> 2 tablespoons raisins
> mayonnaise, lemon juice, dash of salt

Dice the apples, leaving the skin on. Combine with celery, walnut meats, raisins, and salt. Moisten to taste with mayonnaise thinned with 1 teaspoon lemon juice. If you prefer a sweeter taste, omit the lemon juice.

Another apple dish of this genre kindles memories of college dormitory meals when those halls were invested more in gentility than egalitarianism. And in New York's now defunct Automats small servings of this salad lay ensconced in birdbath dishes sunk in ice beds adjacent to the plastic jello cubes. To confess a minor perversion I still find this dish somehow attractive.

APPLE, CARROT AND RAISIN SALAD
for 1

> 1 medium apple, unpeeled
> 1 carrot
> 2 tablespoons raisins
> mayonnaise to taste, lemon juice, dash of salt

Dice the apple, and combine with the grated carrot, raisins and salt. Add mayonnaise and lemon juice.

Apples combine very well with cabbage, raw or cooked. Use some of the outside, greener leaves with the white.

APPLE SLAW

> 3 cups cabbage, thinly sliced or shredded
> 3 tablespoons wine vinegar
> 1 large tart apple, sliced and cut up
> 1 teaspoon caraway or celery seeds
> ½ cup sour cream
> ¼ cup mayonnaise
> salt and pepper to taste

Sprinkle the wine vinegar over the cabbage, mix well and let the mixture stand for an hour. Then cut up the apple, and add with the remaining ingredients to the cabbage. Unlike a green salad this benefits from sitting a while *after* the dressing is on.

A main dish salad that is good at luncheon or supper is:

CHICKEN (OR TURKEY) SALAD WITH APPLES
for 6 to 8

> 4 cups cooked chicken or turkey
> 2 or 3 tart apples, diced
> 1 cup of celery, diced
> 2 tablespoons sweet onion, finely cut
> ½ cup or more mayonnaise
> 1 tablespoon wine vinegar or lemon juice
> 1 teaspoon curry powder
> salt and pepper to taste

Combine all the ingredients lightly but thoroughly. This dish, too, should sit about an hour before serving.

A dish of northern European origins is the redoubtable herring salad, which can be the centerpiece of a smorgasbord or cold table, piled in a great mound and surrounded by sliced, hard-boiled eggs. Even in its simplified form it is rather a lot of work, but it is marvelously satisfying both to eat and behold. Its authentic version requires the soaking of herring, roasting a veal shoulder and suchlike. If you care to go to such lengths any German-accented cookbook will supply the recipe. I find the following recipe a respectable compromise between the genuine article and the pallid delicatessen version:

HERRING SALAD
for 10

>   6 to 8 marinated herring fillets
>   5 medium potatoes, enough to make 4 cups, boiled in the
>   skin and then peeled
>   3 large apples, cored, unpeeled
>   1 cup celery, including leaves
>   3 hard-boiled eggs
>   ½ cup onion
>   ½ cup dill pickles
>   1 cup pickled beets
>   sour cream

All the above ingredients, except the sour cream, should be diced and gently combined in the order given. Dress with the following:

>   ¾ cup wine vinegar
>   ¼ cup sugar
>   4 tablespoons horseradish
>   pepper and salt, which must be tasted for seasoning, since
>   there is no way to anticipate the amount needed

Let the mixture stand several hours and taste again. Incorporate ¼ cup of sour cream if you wish to minimize the acidity, or serve the sour cream as a side dish. Pumpernickel, sweet butter, and beer are proper accompaniments to herring salad.

A final use for raw apples is to end a fall or winter meal with a bowl of one or two varieties, the very best you can find. Handsome as these are, don't leave them as a centerpiece very long, as they need refrigeration. Accompany the apples with a bowl of walnuts and Malaga raisins. If you've had wine with dinner keep it at the table to drink along with the fruit and nuts. Or, more conventionally, bring out a good sherry or port. A classy dessert.

If cooking with apples makes you think of a suckling pig, mouth agape with a ruddy fruit between its tusks, be assured it is not necessary to go to such lengths. While apples and pork do have a special affinity, there are many ways in which the fruit combines with other food in a satisfying manner. Either in the actual cooking process or in the serving at table there is more than simple compatibility; the meat is enhanced considerably, and the apple gains too. Cuisines as disparate as the (Anglicized) East Indian and the German have given us recipes for main dishes with apples. Sometimes the fruit is an almost anonymous addition; in others it is an equal partner. Sometimes it is added at the last minute and holds its shape and identity; at others it may be grated or added in pieces which disintegrate and only remain as flavorful seasoning. Obviously, the variety of apple used would be a tart one in all the following meat recipes. A Rhode Island Greening or Jonathan would be excellent, or any of the tart summer apples, such as Wealthy.

To begin with, a curry, reasonably authentic and good. If you associate homemade curries with leftover desiccated lamb or chicken, try this:

LAMB CURRY
for 4 to 6

> 2 pounds boneless lamb shoulder, cubed
> 1 ½ tablespoons butter or margarine and 1 ½ tablespoons corn oil
> 2 onions
> 3 celery stalks
> 2 garlic cloves, minced
> 2 apples, unpeeled
> 2 tablespoons curry powder (a good Indian import, such as Sun Brand)
> 1 teaspoon turmeric
> ¼ teaspoon cinnamon
> bay leaf
> 3 fresh tomatoes (or 1 cup canned, drained)
> 3 tablespoons raisins

Cut up the onions, celery, and apples, but keep each separate. Sauté the meat, then the celery, onions, and garlic in the oil and margarine, and when the vegetables are lightly colored, add the curry powder, turmeric, cinnamon and bay leaf, stirring the spices into the hot mixture. Then remove as much onion and celery as you can, and hold them aside. Add tomatoes and let mixture simmer for about 50 or 60 minutes, until the lamb is almost tender. If dry, add ½ cup chicken broth. Return the vegetables to the mixture for another 10 minutes, and with 10 minutes remaining add the apples and raisins. Serve, preferably in the pot in which the meal was cooked, accompanied by brown or white rice and a side dish of cucumbers and sliced onions in yogurt. And of course a good chutney, apple if you have it.

The same recipe can be used with beef cubes but must have a longer cooking period for the meat.

The dishes that one thinks of as being typically American rather than derivative generally seem to begin with pork or ham. Old-fashioned and rather countryish is this recipe:

PORK CHOPS AND APPLE RINGS
for 4

> 4 thick chops, loin or rib, trimmed of excess fat
> 2 or 3 tablespoons butter
> 3 onions
> ½ cup chicken broth
> ½ cup apple juice
> salt, pepper, pinch of sage
> 3 apples, cored, unpeeled, sliced horizontally

Brown the chops in butter, and place them in a casserole, or use the frying pan if it has a heat-proof cover. Place the onion slices on the chops, and add the pan drippings, seasonings and liquids. Cover and bake for an hour at 350°. Add the apple rings to the casserole, basting with the pan liquid, and bake another 15 to 20 minutes uncovered. Very nice with baked white or sweet potatoes and a sturdy coleslaw.

When approximately the same ingredients are used for a whole piece of pork the result is a little less cottagey.

ROAST PORK
for 4 to 6

> 1 half pork loin, either side
> Worcestershire sauce
> salt and pepper
> 1 garlic clove
> 2 onions, sliced
> 3 apples, cored, unpeeled, sliced horizontally
> apple juice

Trim a 5-pound or 6-pound roast of all excessive fat, leaving only the thinnest covering. Season with 2 tablespoons of Worcestershire sauce, salt and a good amount of pepper. Insert slivers of garlic under the flesh of the roast. Place the meat in a fairly deep-sided roasting pan, using a rack if you have one, put the onions on top, and roast in a 350° oven for 30 minutes per pound or until no pink runs when you cut into the roast. Baste frequently with the apple juice. During the last 20 minutes arrange the apple rounds on the pork or in the drippings if they're not too fatty. When you serve the roast (on a hot platter) either whole or already sliced, place the apple rings around it as a handsome and edible garnish.

A casserole dish for two, appropriate to the bland modern ham most of us eat, is

HAM SLICE WITH YAMS AND APPLES
for 2

> 2 yams
> 1 ham slice
> 5 tablespoons butter
> ¼ cup brown sugar
> 1 large apple or 2 small ones
> ¼ cup apple juice, chicken broth or white wine

Peel the yams and slice them lengthwise in 1-inch widths. Place the ham in a buttered casserole after slashing the rind in one or two places so the edges will not curl. Put the yams on top of the meat, and dot with butter and sugar, leaving out one third of each. Cover the casserole and bake in a 375° oven about 40 minutes. Remove the cover and test the yams for tenderness. They should be almost soft to the fork; if not return them to the oven. When they are tender add the apple slices and the remaining sugar and butter, baste with liquid and let bake uncovered for 15 minutes. Fresh green beans make a good accompaniment.

When you bake a whole or half ham, use apple juice for a basting liquid. Or apple wine if you can find an acceptable one. You *could* use part Calvados, but that's akin to pouring a bottle of champagne over chicken; I can't bring myself to such uses for great — and expensive — liquor. Self-indulgence has its limitations.

These basic ways of cooking pork and ham are examples of the harmony that exists between anything piggy and a good apple.

You can adapt the recipes above by substituting other pork products — butt, picnic or cottage hams for the roasts; bacon or sausage where the chops or sliced ham is called for. If you use bacon slices or sausage, though, precook them to eliminate excess fat.

All poultry seems to be enhanced by apples. Try a simple stuffing of seasoned, cut-up apples and onions for roast duck or a goose. If your bird is very fat you may not care to eat the stuffing, but it will have served its purpose by flavoring the dish.

With chicken, though, the apple blends with subtlety and strength; both will be the better for the other's ministrations. To begin with the simplest:

BROILED CHICKEN
for 3 or 4

> 1 2½- to 3-pound chicken, split
> 3 tablespoons corn or peanut oil
> ¼ cup apple juice
> 1 tablespoon onion juice
> 1 teaspoon salt
> 1 teaspoon pepper

Combine the liquids and seasonings and rub the mixture well into the chicken. Let it soak, using the extra liquid to turn the chicken in, for at least 4 hours.

Broil the chicken for about 12 to 14 minutes on each side, or until it is brown. Turn off the oven and let the chicken stand for an additional 10 minutes. Rice or noodles and a strong vegetable like broccoli accompany this effectively.

A richer and more elaborate chicken and apple combination,

French in origin, is a fine company dish, so amounts for six are given. It's not too pretentious to call this:

CHICKEN NORMANDY
for 6

 2 small fryers, cut up
 ⅓ cup butter
 ½ cup apple juice
 ½ cup vermouth, very dry sherry or white wine
 2 onions, cut up
 ½ cup heavy cream
 salt and pepper to taste
 3 apples, sliced or cut

Sauté the chicken pieces in butter. Remove them as they brown, and add the onions till they yellow. Place the chicken and onions in a shallow baking dish, and pour the apple juice and the wine into the frying pan. Over a medium flame, scrape up all the brown bits that have adhered to the pan. When the liquid has cooked down somewhat, add the cream, salt and pepper, and continue cooking slowly for about 5 minutes. Pour this sauce over the chicken and onions, add the cut apples and bake, covered, for about 20 minutes. (Use aluminum foil if your pan has no cover.) Remove the cover and bake uncovered for another 10 minutes. The liquid should be saucy but not soupy; add more of the liquids judiciously during the baking process, if needed, and baste occasionally. Serve in the baking dish, along with very fine noodles and honeyed, braised carrots.

There is a category of dishes whose chief charm to their cooks seems to lie in some mysterous ingredient which you are urged to guess. Cakes made with canned tomato soup, meat dishes with sweetened coconut — that sort of thing. The next recipe might be thought of in this way, since beef and apples are not generally associated, but try this excellent version of an old standby:

MEAT LOAF
for 4 to 6

>   2 pounds lean beef, chuck or round
>   1 large apple, peeled
>   1 large onion
>   1 stalk celery, including top leaves
>   2 tablespoons fresh parsley, preferably Italian
>   ¾ cup canned tomatoes, part liquid
>   1 egg
>   1½ teaspoons salt
>   1 teaspoon pepper

Grate the apple and onion, cut up the celery and parsley and mash the tomatoes. Combine all the above ingredients, and shape them into a loaf in an open pan. Bake in a 350° oven for 1¼ to 1½ hours.

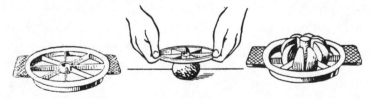

In a beef pot roast or stew, apples are equally anonymous but effective. Simply toss in one large apple or two small, peeled and cut-up tart apples when you add the liquid to either dish (after the beef is browned and seasoned). The apple will disintegrate and help make a lovely gravy.

There are several ways that apples combine with other vegetables to make good main course dishes. One of these is the ham dish given above, without the ham.

APPLES AND YAMS
for 4

    ¼ cup butter
    ¼ cup brown sugar
    2 tablespoons apple juice
    1 teaspoon salt
    3 large apples, in 1½" slices, cored, unpeeled
    3 large yams, cooked, peeled, sliced horizontally 1" thick

Make a syrup of the butter, sugar, juice and salt by simmering them together in a small saucepan for about 10 minutes. In a buttered baking dish place alternating layers of apples and yams, drizzling the syrup over them. Bake covered at 350° for 30 minutes and uncovered for another 15. If you want your dish to taste more like candied sweet potatoes or yams, increase the amount of sugar and butter by half and bake another 10 minutes.

A dish with a somewhat similar flavor, but a different starting point, is stuffed winter squash. Use any of the smaller varieties, such as acorn, butternut or (the very best) buttercup. These are the turban-shaped ones with a little topknot.

STUFFED SQUASH
for 4

    2 small winter squash
    ⅓ cup apple juice
    2 apples, peeled, cored and cut in cubes, or 1½ cups
        applesauce
    ¼ cup brown sugar
    4 tablespoons butter
    ½ teaspoon salt

Split the squash in half carefully. A mallet helps the knife along.

Remove the seeds and pulp. (If you like to feed the birds, save them to put out with your usual seeds. Simply discard pulp and place the squash seeds in a slow oven for half an hour to dry out, while you're using the oven.) To get back to the squash: place them cut side down in a shallow baking pan that has been covered with aluminum foil. Put ⅓ cup apple juice in the pan and let the squash steam and soften for half an hour at 350°. Then turn the squash right side up, fill the cavities with the rest of the ingredients and bake uncovered in the same pan for another 45 minutes or until the squash seems quite tender to a fork jab. If you like a glaze, add an extra bit of brown sugar on top for this "upright" baking.

A vegetable combination with quite another kind of flavor is the Germanic dish of red cabbage with apples. This has a sort of sweet and sour heartiness that fits with a beef pot roast and potato pancakes. It calls for a cast-iron constitution, but it's *good*.

RED CABBAGE AND APPLE
for 4 to 6

> 2 bacon strips
> 1 medium onion, cut up
> ½ head of large red cabbage
> 2 apples, sliced, peeled
> ¼ cup apple vinegar
> 3 tablespoons brown sugar
> 1 teaspoon salt, ½ teaspoon pepper, or more, to taste

Fry the bacon strips, removing the pieces when crisp. Reserve the bacon for crumbling on top when the dish is served. Sauté the onion in the bacon fat until it is yellow. Remove the hard core of the cabbage and cut the remainder into slices, or shred it in a coarse grater. Put the cabbage in a heavy pan with just enough water to prevent sticking, cover it and let it steam for about 15 minutes. Then add all the other ingredients, including the onion and bacon fat, and stir. Cook, still covered, for a half-hour. The flame throughout should be very low, with the cooking time quite a bit longer than is nutritionally optimal, but the flavors do need time to combine. The original version of this recipe had the cabbage simmering on the back of the stove for around two to three hours!

This recipe can also be made with green cabbage, but the flavor is less pungent.

The Union Apple-paring Machine
is so constructed that
The Knife Pares going
both ways,
Thus saving time without increasing the speed of the apple
—o—

Another interesting vegetable dish is

TURNIPS AND APPLES

 a few turnips, white or yellow
 ½ as much apple, peeled, cut up
 butter
 pepper
 1 teaspoon caraway seeds

Peel and cube the turnips, simmer them in salted water to cover and when they are about ¾ done add the apples. When the turnips are tender, drain off the water, add a goodly amount of butter, a grind of pepper and the caraway seeds. Then cook for 5 minutes longer over low heat.

Here are a couple of soups that are improved by the addition of apples. One is

CHICKEN CURRY SOUP
for 4

>   2 medium onions, cut up
>   2 celery stalks, cut up
>   3 tablespoons butter
>   1 ½ tablespoons curry powder
>   1 quart chicken stock, homemade or canned
>   2 medium apples, peeled, cored, cut up

In a large pot, sauté the onions and celery in butter until they are limp, adding the curry powder at the end. After 1 or 2 minutes, stir in the chicken broth and add the apples. Let the mixture simmer uncovered for about 30 minutes. Then, a cup at a time, whirl the mixture in your blender until the whole is combined and slightly thickened. If you don't have a blender, use a food mill, or mash it all through a sieve. A piece of leftover chicken, diced, to equal ½ cup, is very nice as a garnish. Or, if you like a thicker soup, put the chicken in before blending. Season after tasting with salt and freshly ground pepper.

The other soup is a Polish-style cabbage-beef-tomato:

EVERYTHING SOUP
for 6, more or less

> a couple of soup bones, one of which should
>   contain marrow
> 1½ pounds or more beef — brisket, chuck,
>   plate, or whatever
> 1 No. 2½ can whole tomatoes, *not* with purée
> ½ head green cabbage
> 3 onions
> 2 large apples, peeled, cut up small
> 1 garlic clove, whole
> 1 lemon slice
> 2 tablespoons brown sugar
> bay leaf
> salt and pepper to taste

This recipe is cooked in sequence according to the durability of the ingredients. Start with the bones and beef, and barely cover them with water. Simmer them for about 2 hours, skim, then add the tomatoes and cabbage, and cook another hour; next come the onions, apples, and all the seasonings, and another 45 minutes of slow cooking. Do a lot of tasting for seasoning, retrieve the bay leaf, garlic, and bones, and serve the soup piping hot with, perhaps, a dollop of cultured sour cream with minced dill. For a whole-meal-dish place a slice of meat in each soup bowl before you ladle out the broth and cabbage. Serve a good strong bread, such as sour rye or pumpernickel. A fine Sunday night supper.

That really disposes of soups with apples as far as I'm concerned. In Scandinavian cookery, one finds fruit soups, but the thought of beginning a meal with a sweet liquid is not to my taste. If you regard these soups as stewed fruit compotes, reduce the amount of water, and add some wine and spices. But I do think they belong at the end rather than the beginning of dinner.

Stewed fruit, which has such a pedestrian sound, can indeed be a very satisfying dessert, and apples alone or in combination, while simple, are good. Choose a variety that does not disintegrate easily, as does the McIntosh. Jonathan, Cortland, or Golden Delicious are good.

POACHED APPLE SLICES
for 4

> 2 cups water
> ½ cup sugar
> 4 peeled apples, cored, cut in 8 crescents each
> 2 lemon slices
> 2 orange slices
> 1 tablespoon fruit-based liqueur

Simmer the water and sugar together for about 30 minutes, then slide the apple crescents into the syrup, a few at a time so they won't discolor. Let the crescents cook slowly until a fork can just pierce them. Remove them with a slotted spoon, and cook the rest the same way. When all are done, but not mushy, pour the remainder of the syrup and an orangey liqueur, such as Cointreau, over the apples. Serve well chilled. Elegant.

And applesauce, that most basic of stewed fruits, should be given proper consideration. For a very small work investment the result is immeasurably better than the canned product. This assumes starting with a good apple, preferably two varieties.

There are two basic ways of making applesauce. One requires

peeling the fruit first, in which case you lose the nutrition and taste of the skin but which results in a superior texture, one which has not been homogenized by passing through a sieve or food mill. The other method means coring and quartering but not peeling the apples, letting them cook down to the desired soft, not liquid, state and putting everything through a food mill or sieve.

In either method the precepts are the same: Use the heaviest pan you have, use a tablespoon or two of apple juice or water to prevent sticking, use part McIntosh if available, use a minimum of sugar or none at all. If your apples are not as flavorful as you would like, add a slice of lemon, rind and all, a dash of cinnamon and a taste of sugar to balance the added tartness. Good applesauce not only serves as dessert and a main-dish accompaniment, but it can also be the base of many other desserts. It freezes with no loss of flavor or texture.

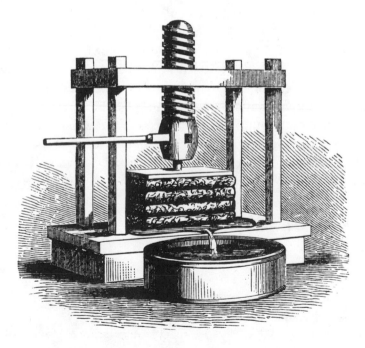

And, staying with simplicity, have you considered the sturdy yet versatile baked apple? It can be plain, or considerably gussied up, and in Europe it is found on quite elegant menus in its more fanciful versions.

For a perfect baked apple the skin must remain unbroken, with no oozing pulp. There is a foolproof method to prevent disintegration, similar to that used in the squash recipe. But first select an appropriate variety. In restaurants an apple is almost always big enough for two, and usually only the largest varieties are chosen. Rome Beauty has become the staple baker because of its size, firmness of flesh and handsomeness. But for flavor there are superior varieties. If you come across an old-timer, the 20-Ounce Sweet, try it, but you're more likely to find a Cortland, Northern Spy and in some areas a York, all of which are good choices. Not everyone wants a six-inch-across apple facing them, after all. But the Rome Beauty is the one you'll probably have available.

As to method: All the following recipes for baking require this first step. Core the apple, peel the stem end with a potato peeler, down about 1½". Place the fruit *peeled end down* in a baking dish which is close in size to the number of apples being baked. Pour apple juice or water over them to the depth of 1 inch. Cover the dish, using aluminum foil if it has no top. If the apples are large, bake them in a preheated 350° oven for about 45 minutes, less if smaller. You then are free to try any version. The most basic is

PLAIN BAKED APPLES

    4 apples
    4 tablespoons butter
    4 tablespoons sugar, brown or white
    1 teaspoon cinnamon
    3 tablespoons raisins

Invert the apples after completing Step 1. Butter the top and as much of the cored portion as you can. Combine the sugar, cinnamon and raisins, and stuff them into the cavities, keeping the raisins away from the top to avoid burning. Return to the oven, uncovered, and bake for about another half-hour, depending on the size and variety. Baste the apples with the liquid in the pan which has begun to thicken and become syrupy; add more juice if necessary. Ten minutes before the apples are done, sprinkle a little extra sugar on the top for a glaze if you wish it. Serve hot or cold.

Some Variations After Step 1

1. Substitute maple syrup or honey, in somewhat smaller quantity, for the sugar. Add a bit of lemon or orange peel. Bake.

2. Combine ½ cup of apricot preserves (for four apples), 2 t. lemon juice, 2 slices of unpeeled orange *cut up small*, 2 T. apple juice and let simmer for 10 minutes. Then add 2 T. of Cointreau or Kirsch. Stuff the apple cavities with cut-up dates, about three per apple, and then pour the syrup over the apples and let them bake for half-hour, as above. This is a stylish but restrained end to a company dinner.

3. Buttered rum and nutmeg. A winter night's dish, and good dressed up with whipped cream. For 4 apples: 4 t. rum, 3 T. butter and 4 scrapes of whole nutmeg. Spread the apple tops with butter, sprinkle with rum and nutmeg and bake. Serve warm.

A fourth variation should be described separately. It is European, more work, but worth the fuss. Since it is served cold, it can be done ahead of time.

APPLES BAKED WITH
RICE CUSTARD
for 4

Prepare four apples as described in Step 1. Remove from the oven and drain off the apple juice, which will be used for basting later. A well-buttered, deep casserole is needed to place the partly cooked apples in. While they are in the oven, prepare the following:

> ⅓ cup white rice
> 2 cups milk
> ½ teaspoon salt
> 1 egg, beaten
> 2 tablespoons butter
> ½ cup raisins
> 1 teaspoon cinnamon
> ½ teaspoon vanilla or 1 teaspoon Cointreau
> 1 teaspoon grated lemon rind
> ½ cup sugar

Cook the rice in a covered double boiler in the salted milk. It will take the rice close to an hour to absorb the milk and become tender. Then add all the other ingredients carefully so as not to mash down the rice, and fill the apple cavities with the mixture, arranging the remainder around and under the apples. Return the dish to the oven for another 35 to 40 minutes at the same temperature. If the mixture seems dry as it bakes, drizzle some of the reserved apple juice over it, or add extra juice. Serve warm or cold. This is a substantial dish, perhaps oversubstantial for ending most meals but fine for an omelet supper.

Puddings have an old-fashioned aura about them. Dessert for most Americans has come to mean either a store-bought cake or pie, or, for the more restrained, fruit of some sort. Yet most of the apple dishes in this category are neither difficult nor heavy. One that is quite special is Brown Betty, one of many recipes designed for leftovers. All too often these are destined for the garbage can; bread particularly rarely goes stale so chemicalized is the original, but it too is seldom used after a few days. This crisp pudding, steaming with apples, spices and raisins, is well worth starting from scratch with your best bread. And there's the rub. It can be made with ordinary commercial white bread, inflated though it is, but it's very much better when you begin with an honest loaf, homemade or bought. This recipe is one that has grown rather far away from the economical version of our grandmothers' time, which would not have included the butter, raisins or lemon.

BROWN BETTY
for 4

> 5 slices bread — if the bread is not firm, toast it
>   until golden tan, then cut in small cubes
> 4 good-sized apples — McIntosh are a good choice
> ½ cup brown sugar
> ½ cup raisins
> ½ cup butter
> 2 tablespoons apple juice
> juice and grated rind of ½ lemon
> 1 teaspoon cinnamon

Butter a 1½- or 2-quart casserole that has a top. Layer all the ingredients except the apple juice, alternately, sprinkling the sugar and lemon and dotting the butter. End up with a thin layer of bread cubes, rub them in your hand so they become crumbly and sprinkle them over the top. Dot with a little butter, sprinkle the apple juice over the mixture and place the dish, covered, in a 375° oven for a half-hour. Remove the cover and bake an additional half-hour or until the top is nicely brown and the inside has shrunk down and is lusciously soft.

An equally good relative of this is

APPLE CRUNCH
for 4

> 4 well-flavored apples, peeled, cored, sliced
> 1 teaspoon lemon juice
> 1 teaspoon cinnamon
> ¼ cup apple juice
> ⅔ cup flour
> ½ teaspoon salt
> ⅔ cup brown sugar
> ½ cup butter

Place the sliced apples in a well-buttered baking dish or deep-dish pie pan. Flavor with lemon and cinnamon evenly, and sprinkle with apple juice. Blend flour, salt and sugar, and cut in the butter as if you were making piecrust. When the flour and butter are combined in a crumb and all the pieces are about the same size, spread this topping over the apples evenly and bake in a 375° oven for about 40 minutes. Check earlier to make sure it's not getting too brown, in which case turn down the oven to 325°. Very nice indeed served warm, especially if heavy cream is passed with it.

There are many other apple puddings, some of which can remain safely in obscurity — a proper fate for Brown Ben (stale brown bread, apples and hot water baked until soft). A fine more-or-less Scandinavian dessert to try is

SWEDISH APPLE DESSERT
for 4

> 1½ cups graham cracker crumbs
> ¼ to ⅓ cup brown sugar
> ½ cup melted butter
> 2 cups tart applesauce, seasoned with lemon
>   juice and 1 teaspoon cinnamon

Combine the crumbs with sugar and butter till all adhere. Keep aside about ⅓ cup of the mixture and press the remainder against the sides and bottom of a buttered pie plate with a spoon and your fingers. Chill the shell for an hour, then fill it with applesauce. Sprinkle the remainder of crumbs on top and bake in a 375° oven for 20 minutes. Serve at room temperature or chilled.

The same ingredients can be used differently. In a small but deep buttered casserole, layer the applesauce and crumbs, ending up with a crumb topping. This dish goes by the name Maiden's Veil and is served with whipped cream.

The one apple dish that transcends gastronomy to the point of being the inevitable American symbol is, of course, Apple Pie. In earlier days it was encountered at breakfast, dinner and supper, when rural America ate in that order. Today it may be dessert or a beverage accompaniment almost any time, sometimes fancied up with such variations of name as French, Dutch or crumb apple pie. But in whatever form it appears, with or without a slab of good store cheese, it has long since been institutionalized as *the* American dish.

Granting a subjective taste factor — degree of sweetness, the presence of cinnamon or even nutmeg, the hint of lemon, the weighty question of thickening — past all these everyone would agree, surely, as to certain standards: the apples should be some-

what tart, certainly flavorful, moist, tender but not disintegrated;
the crust should be short but not greasy, intact but not tough.
If you have your own recipe to achieve this, fine. If not try this:

APPLE PIE
for 6 or 7

The crust should be made ahead of time so that it can be chilled
for at least 1 hour before it is rolled out. The orthodox American
two-pie crust is

   ½ teaspoon salt
   2½ cups flour, sifted before measuring
   ¾ cup shortening — ½ butter and ½ vegetable
        shortening is good
   ⅓ cup ice water

Mix the salt into the flour. Cut the shortening in, using a pastry
blender, until a fairly uniform crumb is obtained. The texture
will vary from pea size to granular. Then sprinkle in some of
the ice water, stir with a fork and add the remainder of the water,
incorporating it so that no flour particles remain. Gathering it
all up in a piece of wax paper may help. Let it chill for at least
an hour. Divide the mixture into two balls, one somewhat larger
than the other. Roll out the larger one on a pastry cloth, using
a lightly floured rolling pin. Rolling from the middle of the
circle, try to make it as even as possible. When it is thin enough,
place it in the bottom of the pie plate, letting the edges hang
over about ¾″.

While the crust is chilling, prepare the apple filling. As to
variety, my favorite of commercially available apples is Rhode
Island Greenings, a fine fruit. Any number of others such as
Cortland, Northern Spy and York Imperial do well too. Some
even use a McIntosh, which leaves you with a nice flavor if less
sugar is used but a rather soft texture.

3 pounds of apples, peeled, cored, cut in eighths
½ to ¾ cup white sugar
2 tablespoons butter, cut up
1 teaspoon lemon juice or rind
1 tablespoon flour or cornstarch
cinnamon, optional

In a large bowl gently turn the apples as you sprinkle over them all the remaining ingredients. The apple slices should be coated with the sugar mixture and placed in the rolled-out lower crust. It will seem to be too high, but mound up the center, and rely on the apples' shrinking. Then roll out the remaining crust and place on top. Join the two crusts by crimping them with your forefinger and thumb. At any point the pastry may tear; think nothing of it and simply patch it up with another piece or the trimmings after you've rerolled them. Now, sprinkle with 1 T. sugar, and, with a small sharp knife, puncture the crust in one or two places, perhaps by making a small initial A, so that the steam can escape. If the apples are juicy, they might well run over the pan unless you are using a deep-dish pie pan. As a precaution, stick a couple of quills of macaroni through the vents, and/or put aluminum foil on the bottom of the oven to catch the delicious ooze. If it does overflow, and you do catch it on the foil, carefully scrape it up with a spoon and put it back under the crust via the vents.

Sometimes I make that same apple pie without the bottom crust, thereby sacrificing ease of service — it won't cut as nicely — to calories. (If you look at people's plates after they've eaten pie, if anything is left it's almost always the bottom crust.) I then use an oblong glass baking dish, and the servings are square. It's a little harder to roll out an oblong crust, but you can patch a lot.

There is, of course, a really elegant one-crust apple pie, the French *tarte aux pommes*, or *tarte Tatin*. It is made with a richer dough, the one the French call *galette* and the Germans *Mürbeteig,* or short dough. If you are adept at puff pastry, that butter-rich and time-consuming procedure, it too can be used.

APPLE TART, TATIN

All ingredients should be cold, except the butter.

> ¼ teaspoon salt
> 1 tablespoon sugar
> 1 cup and 1 tablespoon flour, sifted
> ⅓ cup butter
> 1 egg yolk
> 1 tablespoon lemon juice
> 1 tablespoon water

Stir the salt and sugar into the flour, then cut the butter in with a pastry blender as in the recipe for regular piecrust. Beat the egg yolk with lemon and water, and work the liquids into the flour mixture with a fork or your fingers. Gather the mixture together in a ball with as little manipulating as possible and chill it well before using.

Meanwhile prepare the apple mixture.

> 5 firm apples, peeled, cut in 6 to 8 crescents each
> ⅓ to ½ cup sugar, depending on the tartness of the apples
> 1 tablespoon lemon juice, or part rind
> ½ cup butter, melted

A shallow glass pie dish is nice for this recipe. Pour about ⅓ of the butter into the dish, and place the apples in an attractive pattern of overlapping slices. Then sprinkle the sugar, lemon and remaining butter on top. Put the rolled-out dough on top of the apples, leaving a steam vent. Make sure to use aluminum foil at the bottom of your oven. Bake in a preheated 450° oven for about 10 minutes, then turn it down to 350° for another half-hour or until the crust is brown. Allow your tart to cool for about 10 minutes, and then turn it out for serving. Have ready a flat plate a little larger in size than the pie plate in which to invert it. Run a spatula or dull knife around the rim, put the serving plate flat against it and gently flip. It's much easier done than said, because while the pie is baking, the butter and sugar caramelize to make a slippery kind of sauce. That's why you have to turn it out while still warm. Of course, it's no great matter if it's simply served from the pan, but it's not as pretty. And when you do turn it out, some of the slices may need rearranging.

A dish that sounds unfashionably "heavy" and is in almost total disrepute is the dumpling. A word should be said in its behalf. Consider: it is made of exactly the same ingredients as pie and is really no more effort. Recipes with two degrees of elegance follow. The first has a German accent, is more imposing — and more work.

APPLE DUMPLINGS, NO. 1

> 4 Rhode Island Greenings, left whole, but cored
>   and peeled
> 3 tablespoons butter
> 1/3 cup apple juice
> 1/3 cup dry sherry
> 1/2 cup sugar
> 1/2 lemon, juice and grated rind
> 1 piecrust recipe, using 2 1/4 cups flour, prepared
>   and chilled

Combine the apple juice, sherry, sugar and seasonings. Pour this mixture over the apples, which you have placed in a small baking pan. Cover them with foil and bake at 350° for 15 minutes.

Meanwhile roll out the piecrust as thin as you can manage without making holes, and cut it into four squares. (There may be extra dough to use for other purposes.) Place each drained apple on a pastry square, with a dab of butter in the core, and wrap each apple, making a sort of package out of the dough. Pierce with a fork at several places to let steam escape. Bake the dumplings in a tin with sides in a preheated 425° oven for 10 minutes, then turn the oven down to 350° and continue for about 40 minutes or until the pastry is a pleasant brown. When that point is reached take the sauce that you prebaked the apples in, pour it over the dumplings and bake an additional 5 minutes. Serve warm with a small amount of sauce spooned over each dumpling.

An easier version is

DUMPLINGS NO. 2

    4 Rhode Island greenings, cored, peeled, cut up
    2 tablespoons butter
    ½ cup brown sugar
    ½ lemon, juice and grated rind
    2 tablespoons apple juice
    1 piecrust recipe, as above, or a shortcake crust

Roll out the dough till it's thin, then divide it into rectangles. Apportion the apples, butter, brown sugar and lemon equally on the dough. Sprinkle them with apple juice, and seal each into a little bundle. Pierce them with a fork for air vents. Bake in a preheated 450° oven for 10 minutes, turn it down to 350° and continue for about 45 minutes or until brown.

There is a whole family of desserts almost as long neglected as dumplings that use a less rich shortcake or scone dough. My favorite is

APPLE COBBLER

> 2 cups flour, sifted with
> 3 teaspoons baking powder and
> 1 teaspoon salt
> ⅓ cup shortening, partly butter
> 1 egg
> milk, about ⅔ cup

Sift the flour, salt and baking powder together. Cut in the shortening. Beat the egg and combine it with enough milk to make ¾ cup of liquid. Add the liquid to the flour, and use a knife to incorporate the ingredients quickly. Place the dough on wax paper and turn it in a light kneading motion a few times, then place it in the refrigerator for about an hour. This is a dough that is very versatile.

Apple Filling

> 4 large apples, peeled, cut rather thin
> ½ cup brown sugar
> ½ cup butter
> ⅓ cup raisins
> 1 teaspoon cinnamon

Combine everything in a bowl, cutting up the butter. When the dough has been well chilled, roll it out in a rectangle about ¼″ thick, twice as long as it is deep. Place the apple mixture on the dough, but do not spread it quite to the edge, as you will be making a kind of roll. From the long side nearest you, carefully roll it over, stuffing the apples back in if they fall out. Place it seam side down in a long cookie pan with sides. Bake in a

preheated 400° oven for 10 minutes and then turn the temperature to 375° and continue baking another half-hour, or until
brown. This should be eaten warm.

A nice variation of the preceding recipe is to proceed as above up to
the point you put the roll in a pan. Instead, slice into 1½ inch
cuts and place them in a prebuttered 9″ x 9″ glass baking pan.
Put in to bake as above, in a 400° oven, but when the 10 minutes
are up, pour over them a syrup made of ½ cup apple juice, 3 T.
butter, ¼ cup brown sugar and a dash of cinnamon. (Let it boil up
for a few minutes before pouring over the slices.) Then continue
to bake for about a half-hour. The results are sticky buns that may
crumble a bit but are awfully good.

Or, for an open-faced dish using the same dough and the same
apple mix, roll out the dough in a rectangle, place it on a baking
sheet with sides and, using your fingers, build up a rim or edging
from the dough. Arrange the apple mixture in rows, pressing
down somewhat. Add sugar, butter, cinnamon and raisins. Place
it in a 400° oven and after 15 minutes turn it down to 350°, and
bake another half-hour, or until brown. The last 10 minutes you
can drizzle about ¼ cup maple syrup over, and continue baking.

And what of cakes? My favorite is made with applesauce, and
when our aged apple trees are bearing and the freezer is bursting
with jars of sauce, I make it with homemade. But several of the
commercial brands are fine for this, to say nothing of being easier.
The end product is moist, spicy, reminiscent of a good dark fruit-

cake but much less rich and heavy, and therefore more versatile. This is a proper cake, incidentally, not a tea bread. Herewith, my

APPLESAUCE CAKE

1 cup sugar, preferably brown, less 1 tablespoon
½ cup butter or margarine
2 eggs
2 cups flour
1½ teaspoons baking soda
1 teaspoon salt
1½ teaspoons cinnamon
¼ teaspoon cloves
½ teaspoon nutmeg
2 tablespoons grated orange peel
1 cup raisins
½ cup walnuts, cut up
½ cup dates, cut up
1½ cups unsweetened applesauce
⅓ cup sherry

Cream the butter and sugar well. Beat in the eggs, one at a time. Sift the flour, soda and spices together, and then very lightly combine them with the sugar-egg batter. Then add the grated peel and sherry with the raisins, nuts and dates. (The two latter may be omitted or increased, depending on your desire for richness.) Finally incorporate the applesauce, but don't really beat it once the flour is added. The batter will be quite thick.

Pour the batter into a paper-lined, greased 9″ x 5″ loaf pan, and bake in a preheated 350° oven for about an hour and a quarter. It's done when it shrinks a little from the sides. Remove it from the oven, let it stand for 10 minutes, and then carefully remove it and let it cool on a rack. This is important to prevent sogginess as the cake cools.

You may serve it plain, or with a sifting of confectioners' sugar on top or with a simple glaze: combine ½ to ¾ cup confectioners' sugar with enough orange juice, milk or cream to spread thinly on top.

Amounts are given for one cake, but the recipe doubles easily and freezes well. The cake will also stay moist and delicious for days, though it should be kept in the refrigerator after two or three days to avoid mold.

A different apple cake is of interest to those who are cholesterol conscious. It has a substantial texture and is a good tea or coffee accompaniment.

BETTY GREGORY'S FRESH APPLE CAKE

In a large bowl, mix

> 3 cups apples, peeled, cut in chunks
> 2 cups sugar (this can be decreased by ½ cup)
> 2 teaspoons cinnamon
> 1 teaspoon nutmeg

In a small mixing bowl, beat till foamy

> 2 eggs, add
> 1 cup corn oil
> 2 teaspoons vanilla

Add to apples

> 3 cups flour
> 1 teaspoon salt
> 2 teaspoons baking soda

Add the egg mixture to the flour and apples and mix well. 1 cup raisins and ½ cup chopped nuts are lightly added last. This cake should be mixed entirely by hand except for the eggs, and only corn oil should be used. Bake it at 350° for 1 hour in a greased and papered 9″ x 13″ baking pan. Make a confectioners' sugar glaze if you like.

A group of cakes made with butter batter poured over apple slices, or vice versa, ranges from a quite rich dough to what was known with utmost gentility as Cottage Pudding. Let us avoid extremes; a very sweet buttery cake doesn't do too well with apples, nor does the strict economy version. This is a nice middle-bear choice:

APPLE (BUTTER, BATTER) CAKE
for 6 or so

> ½ cup sugar
> ¼ cup butter or margarine
> 1 egg
> 1¼ cups sifted flour
> 1 teaspoon baking powder
> ¼ teaspoon salt
> ½ cup milk
> ¾ teaspoon vanilla
> 2 large apples, sliced and peeled
> ½ cup brown sugar
> 1 teaspoon cinnamon
> rind and juice of ½ lemon

Cream the sugar and butter well, and beat in the egg. Combine the flour, baking powder and salt, and add alternately with the milk and vanilla. Pour into a greased 8″-square baking pan. Mix the apples with the sugar, cinnamon, and lemon, and place the mixture carefully on top of the batter. They'll sink, but that's all right. Bake in a preheated 350° oven for a half-hour or until it shrinks somewhat from the sides.

And if you've wondered what hidden secrets lie in the kitchens of the Pennsylvania Dutch, here is

APPLE PANDOWDY

> 3 sliced tart apples, peeled, sliced in 6 to 8 crescents each
> 4 tablespoons molasses
> ¼ teaspoon each of cinnamon and nutmeg
> batter of above cake

Place the apples in a buttered 8″-square baking dish. Drizzle the molasses and other flavorings over them. Bake this for about 20 minutes in a preheated 350° oven. Then pour the batter over it and continue baking for about 30 minutes more. Let the dish rest for about 20 minutes and then, using a spatula, carefully turn it out. Or take the lazy way and serve it warm from the pan.

**The Prize Apple Parer at the
Paris Exposition.**

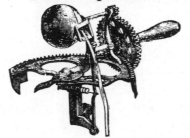

If you bake bread, you probably make a sweet kind, either as coffee cake, stollen or sweet rolls. In case you haven't tried apples with yeast dough, you'll find the combination very rewarding. The yeast process, by the way, is quite easy, requiring a relaxed attitude and some parallel activity that can be interrupted. If you've never attempted such a dish, *The Joy of Cooking* has a good descriptive account of the steps, though like most authorities it advocates more yeast than I find necessary. If you try it you'll be ready for such simple pleasures as

APPLE RING OR KUCHEN

   ½ cup sugar
   ½ cup warm water
   1½ packages yeast
   ½ cup evaporated or plain milk, warmed
   ¾ cup butter, at room temperature
   1 teaspoon salt
   2 eggs
   1 tablespoon orange rind, or part lemon, grated
   4½ cups or more white flour, unbleached preferably

Add 1 t. sugar to ½ cup of warm water, stir and dissolve the yeast in it. Put this aside to rise. Meanwhile, in a large bowl, combine the remainder of the sugar, half a cup of the butter, the salt, eggs and rind. When the yeast-water has risen (about 10 minutes with powdered yeast) add it to the butter mix, and then begin slowly adding the flour. Until this point you can use an electric beater; continue with it until you've incorporated about 2½ cups of flour. Then switch to hand-beating, using a strong silver knife rather than a spoon. Add the remainder of the flour, plus any additional that is needed, until the dough adheres nicely to itself instead of your fingers. Flour a breadboard and knead the dough well for at least 5 minutes, then place it in a greased, clean mixing bowl that is at least double the size of your dough. Cover the bowl with a damp, thin towel and place it in a warm (80° to 85°) protected spot. I use my unlit oven with its pilot light on for a perfect temperature. Let the mixture rise for about an hour and a half, punch it down, add the other ¼ cup of butter and then let it rise again. When it is double in bulk, remove it from the bowl, knead it lightly for a minute and begin to shape as follows:

APPLE RING

Take half the dough, and with a floured rolling pin roll it into a long strip. Or if it's easier, simply shape it with your hands into a ring that will fit into a buttered 9″ tube pan. Peel two apples and cut them into crescent slices. With a small sharp knife make slices in the top of the dough and place the apple slices, pointed side down, in a circular arrangement around the top. Sprinkle it with cinnamon, 2 T. white or brown sugar and dot it with about 3 T. butter. Put the ring back to rise again for about 35 to 40 minutes. Bake in a 350° oven for 35 minutes or until golden brown. Remove it immediately and let it cool on a rack.

APPLE KUCHEN

Using the other half of the dough, after the last rising roll it out into a rectangle about ½ inch in thickness. Place it in a 13″ buttered baking pan. Dot the dough with 2 tablespoons softened butter, then place the apple slices (as above, but use 4 apples) touching each other, in rows. Sprinkle with a little cinnamon, 3 tablespoons white sugar and about another 2 tablespoons of butter. Let the dough rise about 35 minutes, and then bake in a 350° oven for about 35 to 40 minutes.

Enough of baking. There are several recipes formerly known as side dishes and less graphically called relishes today. They don't belong to the grilled steak, baked potato and salad kind of meal but rather to the country table where a variety of taste and texture is offered, testimony to last year's harvest and enterprise. Applesauce finds its place here; so does apple jelly, ketchup, butter and chutney. This last is a treasure and not difficult to do. All the paraphernalia of canning and preserving aren't necessary either. If you want to make about 3 or 4 pints, and you have room to refrigerate the bottles, you can simply cook and store them. They will last for months, but the chances are they won't stay that long.

APPLE CHUTNEY

1½ teaspoons salt
2½ cups brown sugar
2 cups cider vinegar
4 pounds apples, peeled, cored, cut small
2 pounds tomatoes, cut up
3 large onions, cut up
1 14-oz. box raisins
1 lemon, seeded and chopped, skin and all
1 garlic clove
2 tablespoons mustard seed
1 tablespoon chili peppers (the hot flakes used for spaghetti sauce)
Ginger: if fresh, about 3 ounces peeled and chopped; if preserved, about 3 ounces chopped; if powdered (the least satisfactory), 1 teaspoon to 1 tablespoon depending on your preference for the taste

Put the salt and sugar into the vinegar and bring it to a slow boil. Then add all the ingredients and simmer for an hour. Taste for salt and hotness, remembering that it will seem sweeter then than when it has cooled. Correct the seasonings and put the mixture back another few minutes for a blending of flavors. Put your chutney in hot clean jars.

You can vary this by adding a few green tomatoes cut up small if you're left at the end of the season with unripened tomatoes. Substitute them for some of the ripe ones or for an equal amount of apples. And a lime is excellent in place of the lemon.

Here are two recipes that begin with homemade applesauce:

APPLE KETCHUP
one quart, or more

> 8 apples, tart and tasty
> 1 ½ cups chopped onion
> 1 tablespoon salt
> 1 ½ teaspoons cinnamon
> 1 teaspoon dry mustard
> 1 teaspoon ground cloves
> 1 teaspoon pepper
> ¾ cup sugar
> 2 cups cider vinegar

In a large saucepan with a heavy bottom cook about 8 large apples; about a quart of pulp is wanted. Make the sauce, using either method described above, with a minimum of water, but cook it longer than normally; there should be no lumps. Then add all the other ingredients, simmer for 1 hour and put the mixture through a sieve. The fragrance alone will make your reputation as an earth mother.

In 1885 the *American Agriculturist,* surveying the commercial market and the buying habits of consumers generally, found that dried apples were in "large demand for sea voyages, as well as in families." But, they said, "There is also a large demand for apples to manufacture jelly, which is offered for sale in the village grocery stores at very reasonable prices. This article is taking the place of the old-fashioned apple butter or applesauce, which formerly was to be found at almost every farmer's table during the winter and spring months. It still lingers in the rural districts, but the jelly and the canned fruit, now so common, are gradually displacing it."

Today apple butter lingers only as a store-bought and rather gummy substance with all the charm of some low-level fake food, rather than the pungent essence of apple that it is. Needless to say you have to start with a tasty combination of fruits and not rely on spices to mask or supply flavor. Winesap, Cortland, Stayman or Northern Spy make a fine butter in any combination.

APPLE BUTTER

> 3 pounds apples, cored, quartered
> apple juice to prevent sticking
> sugar
> ½ lemon, juice and grated rind
> 1½ teaspoons cinnamon
> ½ teaspoon ground cloves
> ½ teaspoon allspice

Cook the apples slowly in the juice until they are quite soft. Put the fruit through a food mill or sieve, into another pan, measuring the pulp as you do so. For every cupful add ½ cup white or brown sugar. Stir the sugar in, along with the lemon, rind and spices. Then, over a low flame or in a 300° oven, cook or bake this for about 2 hours, or until it is thick enough to mound on a spoon. Stir it occasionally. This amount will yield a little over a quart and won't need processing if it is to be used within a few weeks. It's a pleasant and hearty spread, somewhat less sweet than jelly or jam.

And apple jelly itself is not only delicious but, because of the high pectin content, is useful in combination with low pectin fruits, such as raspberries and blackberries. Below is a simple recipe that you may care to vary with the addition of fresh mint (*don't* add green vegetable coloring) or fresh thyme if the jelly is to be used as a meat accompaniment. Or if you have one of the old-fashioned fragrant geraniums, such as rose or lemon, put a leaf or two into the jar before sealing with paraffin. A remembrance of things past.

BASIC APPLE JELLY

Cook about 4 pounds of cut-up, unpeeled, un-cored apples in 2 quarts of water, for a half-hour, or until tender. If you have a jelly bag, strain the apples through this into another clean utensil. If you don't, use a large square of cheesecloth and allow the juice to drip through. For every cup of liquid, add one cup of sugar, and cook until the sugar is dissolved. Then boil rapidly until a bit jells when placed on a spoon. Skim the froth off the top, pour into hot clean glasses at once, and cover with heated paraffin. Makes about 8 8-ounce glasses.

An easy, more contemporary relish is raw applesauce made in the blender. Peel, cut up small and blend 1 apple with ¼ cup apple juice. Use it quickly or it will discolor.

A relish made with cranberries, orange and apple is also very easy:

THREE-FRUIT RELISH

  2 cups cranberries, washed
  1 Florida orange, seeded
  1 large red apple, cored
  ¾ cup sugar

Put the cranberries, cut-up orange and apple alternately through a food grinder. Add sugar, and let the mixture stand for 1 hour to blend. It is good raw. If brought to a quick boil for a few minutes, it is even better.

Generally, frying is not a mode of food preparation that I'm fond of, but in the case of the next few recipes I forget my misgivings. The first, the puffy apple pancake, is in the style of the old New York Lüchow *Pfannkuchen,* and though it could be a Gargantuan dessert it is really a lunch or late evening supper meal in itself.

APPLE PANCAKE, OMELET OR EVEN SOUFFLE
for 2 or 3

> 2 apples, peeled, sliced — McIntosh, Cortland or a
>   similar flavorful variety is good
> 6 tablespoons butter
> 4 large or extra-large eggs, separated
> ½ teaspoon salt
> 2 tablespoons white sugar

This recipe must be made in a large — 10″ or 12″ — frying pan. Sauté the apples in half the butter until the slices are soft but not brown. Distribute them evenly over the pan, add the remainder of the butter and heat them until they are gently sizzling. Meanwhile beat the egg white with the salt until firm. Put it aside, and, using another bowl but the same beaters, beat the 4 yolks with 2 tablespoons sugar until they are frothy. Then fold in the whites, incorporating all the flecks. Spoon this mixture gently into the hot frying pan with the apples. Put the pan into a preheated 400° oven for about 10 minutes. Sprinkle an additional spoon of sugar on the top before serving. It will collapse in a moment, but the first puff is quite impressive.

The other frying recipe, fritters, is very easy if you own an electric deep-fat fryer, and not really hard if you don't. What's needed, however, is a deep, fairly narrow saucepan. Fritters can be an accompaniment to meat, as they usually are in the South and Midwest, or with very little change they become the elegant dessert fritter, French in origin, served all through Europe and South America. If you've never tried to make them, or even eaten them, don't be put off by what appears to be a certain degree of complexity; they're quite simple, and superb.

FRITTERS
for about 4 or 5

> 4 apples
> 2 tablespoons sherry
> 2 tablespoons confectioners' sugar
> oil
> batter

Peel, core and slice or cut into eighths 4 apples. Try to get a firm-fleshed apple, but any variety that is sweet will serve as a dessert fritter. Sprinkle sherry and powdered sugar on the slices. Let them stand for about 2 hours, then drain them dry and pat them with absorbent towel. This is important, because you want the batter to adhere.

Batter

> 2 eggs, separated
> pinch of salt
> 1 tablespoon melted butter
> 2 tablespoons sugar
> 1 tablespoon fruit-based liqueur, lemon juice,
>   or sherry
> 1 cup flour
> ⅔ cup liquid, all milk or half water

Beat the egg whites and salt until they are stiff. With the same beater, but in a separate bowl, beat the egg yolks until they are light and lemon-colored; then add the butter, sugar and liqueur. Combine the flour and the egg yolks with a minimum of mixing, then add in the milk to make a general-purpose sweet batter. Simply eliminate the sugar and liqueur if the fritter is to be a meat accompaniment.

Take about half the apples and dip them in the batter, letting the excess drain off. The fat should be heating meanwhile to a temperature of about 370°. If you don't have a thermostat or thermometer the classic way is to put a stale cube of bread in the fat. If it is brown in 60 seconds the temperature is about 375°. Make sure your pan is deep enough, oil to about half the depth of your pan. Fry one third of the fritters at a time, otherwise the temperature will lower. The actual frying process takes only a couple of minutes, so be attentive and ready with paper towel and a slotted spoon to remove the fritters as they brown. Put the paper on a heat-proof plate, and then keep the fritters warm in a 250° oven till the final batch is ready. Be sure to let the fat reach its correct temperature between batches. To serve, sprinkle the fritters with sugar. If you want to gild the lily, make a sauce of puréed apricots simmered with a little sugar and flavored with Cointreau.

For pancakes and waffles add a small grated apple to your usual batter, and perhaps instead of syrup sprinkle a combination of brown sugar and cinnamon.

Here is one stylish dessert that is almost as easy as putting out a bowl of fruit. It also has the great virtue of taking only a few minutes to make, so it can serve for those times you completely forgot any sweet is needed.

BROILED APPLE SLICES
for 4

> 4 apples, unpeeled, cored
> ⅓ cup brown sugar
> 1 tablespoon lemon juice
> 2 tablespoons rum or liqueur, optional
> 4 tablespoons butter

Slice the apples horizontally, about ¾ to 1 inch thick. Place them on a shallow pan that can be exposed to the broiler. Use 1 tablespoon butter to grease the pan, and dot the remainder on the sliced apples, sprinkling the sugar, juice and rum on top. Place under a preheated broiler and watch closely. In 4 or 5 minutes they should be bubbly, brown and glazed. Don't attempt to turn them. Serve immediately.

PLUCKING THE APPLES

And what is more melancholy than the old apple-trees that linger about the spot where once stood a homestead, but where there is now only a ruined chimney rising out of a grassy and weed-grown cellar?

Nathaniel Hawthorne,
Mosses from an Old Manse

Coda: The Old Trees

IN BACK OF the house, across the little mint-clogged stream, is a gentle slope where the old orchard stands. The trees are gnarled and gray with scale. Seventy or eighty years ago, when the farmer put the first of the present trees in the ground, the place was a working farm, with meadows for hay and clearings for the grazing cattle. It is likely that an orchard stood there even years before that, judging from the stone walls that mark its boundaries. (Before barbed wire came east from the cattle range in the 1880s, the stones yielded up in the original clearing of the woods were used to build the stone fences typical of this part of New England.)

The existing trees, about 250 of them, were planted in three blocks, separated by two strips of meadow. The bulk of the trees are Baldwin, and because they are very old their fruit often has the tiny brown fleck distinctive of the variety. The large proportion of Baldwin among the present trees is not representative of the original planting because their longevity is far greater than other kinds. There are Winesaps, some Northern Spy, and Golden and Roxbury Russets, all of these testimony to the apple taste of, say, 1900. But in the center of the orchard there may have been a couple of dozen Red Delicious, judging from the empty spaces in the row and the remaining four shriveled, gnomelike trees, each bearing a few poor fruit with the typical Delicious conical shape. These might have been planted around World War I, when the Red Delicious was a trendy variety. What persuaded the farmer to try the unlikely Delicious in our blasting New England winters?

To the left of the house where the land slopes up to the pond a half-dozen or so Pound Sweets barely hold on to life. And fronting the road stand two battered McIntosh, clearly later additions. A couple of Russets span the distance between the two other varieties. Oddly enough, among all the trees there are no summer or early apples, no Astrachan or Wealthy to make into the first pies or sauce. Perhaps these varieties had not the strength to survive the neglect the trees suffered through the years.

At the beginning of the century the orchard was part of the prosperous life of the farm. Rooms were added to the house, and around 1910 a fine new barn was built on the old stone foundation. The family's descendants speak of those days with pride. The apples were in their prime and must have been a handsome sight, as well as a substantial addition to the farm resources. The orchard must have borne every other year, in the way of predominantly Baldwin trees, and persevered, despite a growing disinterest on the part of the farmer, as it became less profitable. By the 1930s only first-class apples were bringing in enough to pay for picking them, and by that time the orchard was of little importance in the economy of the farm. When the place was sold to a prosperous New Yorker and his family it was at first a summer house, and then, in the way that Vermont homesteads have, it assumed a more important place in their lives. Later, a son and daughter-in-law took to the land permanently, and the barn housed chickens, and eggs were sent to market. Still the apple trees continued to bear, but the ground must have been thick with fallen apples. Neighbors came with bushel baskets, barely making a dent in the crop. Again the place was sold, and a disinterested real-estate broker held title for some few years. By this time the orchard was pretty much in the public domain and was known to hunters who waited, in season, for the deer to come out of the woods for the apple falls.

One late September day a few years back we walked behind the not very prepossessing house, over the little stream, keeping up with the fast-talking real-estate agent. There lay the orchard in the still warm sun. The meadow was high and golden with tall grass and deeper burnished goldenrod forming a frame for the rows of trees, heavy with green leaves and bent red with apples that year. The matter was settled then and there; we were to own the place, streams, house, barn, woods, and orchard. Never since have the trees borne so much as one-third the fruit they had that year. It was as if they put forth some seductive trail just for us, knowing that we would succumb.

If such anthropomorphic nonsense came to mind it was not to linger; it is hardly "our" orchard despite attempts to make it respond. We gave ourselves a present of the services of a professional pruner, whose crew, accustomed to dealing with commercially viable trees, looked at these brittle limbs with disbelief. The work went forward. Pruned, then fed and mulched, they were given care appropriate to the aged. The trees "look" better, the dead branches piled in corners of the orchard providing living space for all sorts of creatures. But as to fruit, that seems to be a matter over which we have little influence. Some years are good, and some years aren't, even though they're supposed to be bearing years. Ah well, there are apples enough for our needs, and the neighbors, and the deer, and certainly the mice, and even the unwelcome porcupine.

Midway up the ridge is a stone wall which once separated the orchard from an adjacent field. When the field was abandoned, the stone wall marked the line between the orchard and the scrubby new growth that began the process of the return to forest. By now there is a pretty good mix: fern, wild flowers, berries, with some pine and maple and birch crowding out the skinny poplar. Some scores of trees escaped to the sun and openness of the orchard and established themselves in the once stately rows of Baldwins during the years when the hay was not mowed. All the common wild flowers that grow in sunlight compete here too, and before the annual cutting there are black-eyed Susans, daisies, self-heal, and a dozen more. Closer to earth the red wild strawberry leaves meander through the undergrowth, hugging the ground, their berries few and startlingly delicious, a kind of Platonic ideal of strawberry taste. They are mouse-run high, and our mice dine well.

The orchard is home to a family of fox, numbers of woodchucks, and countless birds, both those who prefer to nest in the knotholes of decaying trees — wrens and woodpeckers and titmice, and the branch nest-builders as well — the robins and goldfinches and phoebes.

There is, too, a whole range of life that is not visible to the eye but is felt only through its works. When we see the tiniest of nips in the fallen apples, or, sadly, note old girdling marks on the bark, this is evidence of the mouse population. In the scab and other ailments of fruit and leaf we see the results of the many viral forms that use the apple tree as host. Codling moth and mite are also the beneficiaries of neglect, benign and otherwise. We will not use chemicals on these old trees, partly out of economic concerns and partly from a confusion of sentimental and "ecological" ones. (Totally inconsistent, we are spraying a half-dozen young trees that we set out a few years ago in another part of the place.)

At the other end of the life scale, the largest mammal to come to the orchard is the fine, many-antlered deer, who stalks in and out at will, if he has survived the hunting season. In the winter, when he leads his harem of does and the youngsters born the previous season, they will trample the heavy snow to unearth the apples lying beneath. These have been frozen and thawed, and on a warm day sometimes there is the aroma of the fermenting fruit, a kind of natural cider.

When the long winter is over, the lengthening days slowly and erratically warm the air and then the earth. Dormant now for many months, will the old trees stir to life? Here and there as we walk among them in early spring we see the damage of snow and wind on the weakest, a limb down, or a whole tree split. Yet somehow even those whose trunks seem almost hollowed out find a source of life, and put forth the first green leaves. And by June the whole cycle of life will be in full swing again: flowers, and then fruit.

> Here's to thee, old apple tree,
> Whence thou mayst bud, and whence thou mayst blow;
> And whence thou mayst bear apples enow.

General Index

Index of Apple Recipes

General Index

Index of Apple Recipes